Effective Speaking
Communicating in Speech

D0702314

JOIN US ON THE INTERNET VIA WWW, GOPHER, FTP OR EMAIL:

WWW: http://www.thomson.com
GOPHER: gopher.thomson.com
FTP: ftp.thomson.com
EMAIL: findit@kiosk.thomson.com

A service of I(T)P®

For David, Sarah and Anna

Effective Speaking:
Communicating in Speech

Christopher Turk
Lecturer in English
University of Wales Institute
of Science and Technology
Visiting Fellow, Yale University, 1983/84

E & FN SPON
An Imprint of Chapman & Hall
London · Weinheim · New York · Tokyo · Melbourne · Madras

**Published by Chapman & Hall, 2-6 Boundary Row,
London SE1 8HN, UK**

Chapman & Hall, 2-6 Boundary Row, London SE1 8HN, UK

Chapman & Hall GmbH, Pappelallee 3, 69469 Weinheim, Germany

Chapman & Hall, 115 Fifth Avenue, New York NY 10003, USA

Chapman & Hall Japan, Thomson Publishing Japan, Hirakawacho
Nemoto Building, 6F, 1-7-11 Hirakawa-cho, Chiyoda-ku, Tokyo 102,
Japan

Chapman & Hall Australia, Thomas Nelson Australia, 102 Dodds
Street, South Melbourne, Victoria 3205, Australia

Chapman & Hall India, R. Seshadri, 32 Second Main Road, CIT East,
Madras 600 035, India

First edition 1985
Reprinted 1987, 1991, 1992, 1994, 1996, 1997

© 1985 Christopher Turk

Printed in Great Britain by J. W. Arrowsmith Ltd, Bristol

ISBN 0 419 13020 9(HB) 0419 13130 6(PB)

Contents

Acknowledgements

Apart from the many people whose talks I have listened to, criticized, and learnt from, during a decade of teaching effective speaking to commercial, industrial and governmental organizations (as well as in several universities) I owe special thanks to several colleagues and friends.

Firstly, John Kirkman, who introduced me to the scientific study of communication, and who nursed me through my own learning process. John and I wrote *Effective Writing* together in 1980/81, and when we delivered the manuscript, I asked him if he would continue to collaborate on *Effective Speaking*. He had other projects at the time, and I went on to write *Effective Speaking* by myself. But John is present in ideas, if not acknowledged on the cover. Although general knowledge about speaking is widespread, and ideas could (and have) come from the many other books on the subject, John was my teacher and friend, and this is my chance to thank him for his help. He taught me about professionalism when I was a young man, and supported me through many early failures.

I also want to thank my other senior colleague, Alban Levy. One of the best speakers I know, Alban combines a sure wit with a masterly knowledge of audiences and their psychology. Just watching him perform was an education in itself; and audiences throughout the world treasure his humour. I have learned a great deal from him, and in many cases have adapted his way of organizing a topic, rather than use ideas from the extensive literature on the subject.

The third colleague is Peter Hunt. Several of the examples are his, and many informal conversations with him have refined my ideas on speaking. Peter is one of those people, intelligent, urbane, and witty, who is a naturally good speaker. My thanks are due to him as a good friend, as well as a colleague, and an example over a decade and a half.

Finally, I would like to thank everyone else who put up with my distractions while writing. Dennis Bratchell, at UWIST, helped me to obtain a sabbatical during 1983/84 when there was finally time to finish the book. My father read drafts and proofs. Without Catherine, though, it would never have been finished.

1

Communicating in speech

Who is this book for?

Everyone has to be a listener sometimes; at lectures, presentations, meetings, and on the telephone, we spend a lot of our time listening to others talking. We all know, then, that the average standard of spoken presentation is poor. We are often bored, irritated, even embarrassed as listeners; rarely are we captivated, or filled with new enthusiasm for a topic and with respect for the speaker. I suppose most people when they suffer an appalling presentation vow that when their turn comes they will do better.

When we are asked to speak it is often not so easy. I expect that many people who pick up this book have either just been asked to give an important presentation, or have just given one which has gone less than triumphantly. Take comfort; speaking, like most of the things we do, can be learned. It is not a mysterious gift, something inborn in the lucky and denied to ordinary mortals. It is a skill, and thinking about it will improve it. This book is written on the premise that careful consideration of the problems which face the inexperienced speaker will improve the standards of presentation they can hope to achieve. The experienced speaker, too, can improve. Habits formed without thought, mannerisms which have been reinforced over years of use, assumptions which have never been analysed, can be remodelled by thinking about the task of effective speaking.

This book is not intended, or expected, to produce demagogues. It is primarily for people who need to speak as part of their job, and whose careers will be advanced by the ability to speak competently. Increasingly, industry and government organizations prefer verbal presentations. Reports of research work, proposals for administrative innovations, progress meetings, union meetings, the training of new recruits, conferences and symposia are just some examples. Indeed, many senior managers find that a case made face-to-face is more effective that one made in reports and memoranda. So most professionals (whether administrators, engineers, scientists, technologists, or those climbing the management ladder), will sometimes find themselves asked to give verbal presentations.

In private life too, people are called on to speak who may be unwilling or ill-prepared. This book is intended to help such people; politicians, advocates, salesmen and revolutionaries are not the readers I expect, although they may (indeed will) find useful ideas. The chapters on intonation control (eight) and non-verbal communication (nine), for example, report research on the psychological impact of speech and behaviour patterns on audiences. Indeed, observation suggests that many supposedly experienced speakers could get a great deal of benefit from this information. This book is written so that it can be used, both for general reading, and as a text book for courses on communication skills. The material in the references can be browsed by the general reader, or studied by the student. Both should be able to speak better afterwards, no matter how deep their study.

Learn to speak well

Most people think a decent standard of competence in speaking to a group is part of the basic professionalism of any job; but too many professionals are nervous about speaking, and afraid that they do not speak well. The basic premise of this book, as I have said, is that such a decent standard can be learned, and this confidence is based on many years of experience in training people to speak. A first stage in building up the confidence to speak is to think about the job of speaking, what tools you will use, and what effects you aim to achieve. Language is the basic tool, and language is a mysterious phenomenon. Consider, for a moment, the basic skills in communicating that everyone possesses. Language is used by all human beings; we use it copiously and without second thought every day of our lives. Indeed, our ability and confidence in manipulating language is a central part of the personality we present to those around us. But there is nothing unalterable about these abilities.

Because language skills grew without conscious thought we imagine that our level of competence in the use of language is something unalterable. If we are hesitant, slow, unimaginative, and pedestrian we fear that this situation is foredoomed, that we cannot change what we are. But our language skills are not what we are. Because we negotiate most of our interactions with the outside world through language these skills may appear to others to represent all that we are, and we often allow ourselves to believe their view of us. But we should not. Language skills can be modified and improved by repeating the same processes we first used to acquire them. We learned our language by listening to others, and imitating what they did. We can improve our command of it in the same way; by listening to research findings and advice based on them, imitating techniques which we observe to work, and thinking.[1]

Many people argue that speaking well is no more than the application of common sense. But in this book the results of research in psychology and linguistics are used to support advice on effective techniques. By doing this, I aim to help the speaker become more aware of the complex interactions between speaker, message and audience. The application of thought to any activity requires us to understand it first.[2] The common sense school will reject all such ideas on the grounds that they are either obvious, or incomprehensible. Such a Luddite approach should be foreign to an engineer or scientist, but surprisingly it is often people who apply rational thought in their jobs, who consider language skills to be in the realm of witchcraft.

I am not unaware of the fears that this approach may evoke. And I am certainly aware that not all research is useful, comprehensible, or relevant. I am not about to bombard you with a textbook of academic psychology. It is certainly true, for instance, that all too often the so-called 'discoveries' of the human sciences are rather obvious. One investigator warns his readers that:

> Occasionally we make discoveries which fail to set the world alight with surprise and admiration. One of the more profound insights we have achieved since 1945 has been the realisation that man speaks. He also listens to speech. Some men read and write as well.[3]

But just because some research confirms the obvious, it should not necessarily be ignored. It is surprising how often speakers fail to use common sense, and surprising how often the obvious has to be repeated when training people to speak. I argue that we should pay attention to the objective, scientific, evidence about how people speak, and how they affect the audience. I agree we should avoid that excessive faith in laboratory responses which George Miller calls 'psycholatry'.[4] We should strike a balance between ignorance of research, and pseudo-scientific over-respect for jargon. Our aim, we should not forget, is to speak better: it is not to become armchair psychologists, nor is it to give up the task as a hopeless case, to which no rules apply, and in which no knowledge can help.

So far I have discussed the role of research findings in developing practical skills as if research on people's behaviour is always a bland confirmation of familiar patterns. But sometimes the results are unexpected (as some of the research reported in this book is – for example the effect of hesitations reported in chapter six.) The most interesting and useful results of psychological research are often the ones which are 'counter-intuitive, that is to say surprising and quite unexpected.'[5] A well-known worker on social psychology, Michael Argyle, reminds us that his subject is:

> full of surprises because many of the research findings could not have been anticipated by a thoughtful person sitting in an armchair and analysing what happens when people meet.[6]

By talking about these interactions more precisely than is common, I hope to make the speaker perform better. But this can only be done if we bring into consciousness skills which are usually unconscious. Many people fear that this will make them lose spontaneity and become painfully self-conscious. But that does not seem to happen. The increased consciousness of what we are doing is usually not noticed by others, while the improved skills most certainly are. Everyone who has been involved in training social and performance skills agrees that thinking about these skills improves them.

If the first premise of this book, then, is that speaking skills can be improved by thinking about them, the second one is that psychology and linguistics have much to teach us.[7] I must be careful, though, that the book is not regarded as a manual for experts, or that I am regarded as a super-speaker who does not understand the fears of the ordinary trembling mortal. I am not. Nisbet suggests that in speaking, as in other things, there are:

> three levels of proficiency: the provisional licence holder, the ordinary road user and the rally driver. The first is learning the rules; the second has everyday skills, and some bad habits; the third can break many of the elementary rules – a dangerous style, but a delight to the connoisseur.[8]

I am aiming to help the ordinary road user. I am not a rally driver myself, and have usually come a cropper when I have tried to be one. It is the decent standard of ordinary competence in speaking that this book aims at.

What needs to be done

If we are to improve speaking skills, we must first become more aware of ourselves, our motivations, behaviour patterns, and likely mistakes.[9] Second, we must be aware of the audience's psychology, and their reactions to the speaker's faults and omissions. The first problem for all speakers is being aware of themselves, and judging correctly their own part in what is, for many, an unfamiliar interaction. Quite a bit of the advice and discussion throughout the book will be about how we achieve this useful self-knowledge. One of the difficulties, for example, is that although we are always trying to present ourselves in a favourable light to others, we have little real idea of what we sound like to them.

The main effect we have is created by the tone of our own voice. Indeed, some people are said to be very fond of this sound! But the sound we hear ourselves is very different from the sound that everyone else hears, because we hear it in a different way. Other people hear us (and we hear other people) only through sound waves in the air. But we hear our own voice mainly as the vibrations transmitted from the voice box, through the bones of the head. Only by trying shouting into a skull from a medical student's skeleton can we judge what a difference these bone resonances make. You can perhaps appreciate what a difference this method of transmission makes by considering how often people are surprised by tape-recordings of their own voices. Psychologists have discovered that we are typically quite unaware of the emotive affects of the way we speak. We may not realize how cross we sound, for example, or how often we interrupt other people.[10]

The second area for careful thought is diagnosing what has gone wrong when a talk fails to have a good effect. The reasons are usually in part lack of knowledge about the audience's perceptions and expectations, and in part general disorganization. In my experience, presentations are often ineffective either because of ill-thought-out behaviour, and lack of confidence,[11] or because of a failure to organize ideas and information in an easily understood way. There is a great deal of knowledge and experience about why talks fail, and this book suggests ways in which you can avoid failure. So take heart; if a talk you have just given has collapsed into disaster, there is hope. The reasons for such failures are fairly well known. If the thought of speaking fills you with cold despair, or even if you are just not very satisfied with your performances so far, there are plenty of solutions. The first problem is having the courage to recognize your mistakes and thoughtlessness. The second (and easier) problem is to correct them.

I shall be following a fairly detailed line through these problems in this book, discussing first the audience, and in the next chapter the analysis, selection and organization of the material. There are then separate chapters on practical aspects of speaking, such as using notes, coping with nerves, and getting the timing of the talk right. The following three chapters deal with making the presentation as varied as possible, being aware of non-verbal signals from both speaker and audience, and arranging the room you are speaking in. Three more chapters deal with specific techniques; the use of visual aids, giving persuasive talks, and handling questions. The reader should then be much more knowledgeable and confident about what he or she is doing. By the end of the book your speaking skills are likely to improve.

If you must give a speech tomorrow, and have no time to read the whole book now, the best single piece of advice I can offer you is this. Practice is the best way to learn a skill like speaking, and if practice is to be effective, you need a critic. This is not just my own hunch; it is a well recognized result in research on social skills that:

> A text can provide a coherent background of concepts and principles where these exist . . . or supply knowledge about techniques, but to teach successfully each individual must practise the skill, receiving feedback on his performance, in order to discover his own particular abilities and failings[12]

Practice is vital, but practice by yourself tells you very little. Who ever felt nervous in front of a bathroom mirror? And effective criticism from a spouse is more likely to result in divorce than in better speaking. The best source of criticism is another person, and the best other person is someone you trust, but who has no axe to grind. Find a friend, and ask him or her to sit and listen to you trying out your talk, if possible in the empty room you will use for the actual talk. Believe what they say to you, for your perception of yourself is nothing like as accurate as theirs. Their advice is likely to be the best quick guide to effective speaking you can find. If they, and you, have time to go further on the quest for good speaking, then get him or her to read this book. It aims to give the framework of ideas, evidence, and anecdote for good speaking. But in the end the substance of good speaking is acquired by intelligent awareness while speaking, and only practice will perfect the advice this book offers.

Communication in theory

Having sketched the field we are about to enter, it is time to start looking seriously at the basics of good speaking. In order to have a clear idea of what we are doing, I want first to deal with general principles which apply to communicating whether the medium is speech or writing, before going on in the following chapters to deal with the special problems of spoken communication.

I am not intending to treat the reader to a specialized tome of linguistic theorizing, or what George Miller calls 'a pleasant field trip through some rather exotic psycholinguistic meadows'[13]. But I do hope to give the reader a genuine understanding of the mechanisms employed in giving and listening to speeches. The bedrock of this understanding is best found in a map or model of the communication process. Claude Shannon and Warren Weaver designed a model in 1949[14] which schematizes the process of communicating in its most general form. This model is one of the oldest, and most commonly used in the study of communication.

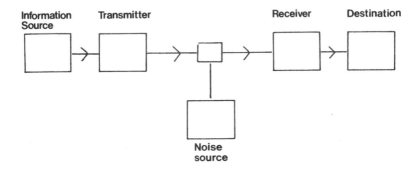

Fig. 1.1 The Shannon and Weaver model

What can we learn from this model? Its virtue is that it helps us to think about the overall process. It highlights some obvious but important aspects of communicating information, which are often overlooked. Its vice is that its simplifications are so extreme that they sometimes seem to trivialize the discussion. Certainly, it should not be taken as the last word on the subject; it is little more than the first. But the model does remind us that the end of the process is not a physical product – *words* or *text* – but a mental result – *understanding*. The main insight this generalized model of communication offers is its emphasis on the flow of information. No communication is just the production of signals, with no thought for their purpose or destination. The writer does not, or should not, see paper as his only end. Similarly, the speaker cannot act as if he or she were talking to an empty room. The listeners are as much part of the process as the speaker. A communication model without an end as well as a beginning, would be like a half built bridge.

What does this teach the communicator? Firstly, that the information must be selected for the needs of the audience, not just for his or her own convenience. He is not just producing information, he has to shape it so that it will fit into the listener's mind. No locksmith makes keys to suit his own ideas of prettiness – they have a job to do. So no communicator assembles his information without a careful eye on where he is to put it. It must fit the recipient's needs, otherwise access to the listener's understanding, memory and approval will be denied. It is a point we would all regard as common sense. And yet it is a point which is most commonly overlooked.

I have heard so many talks fail just because they were over the heads (or beneath the contempt) of their listeners. The needs of the audience must determine the selection of material – it must help them to do their job, or pursue their interests. This is the only ambition in a speaker which will be rewarded with success. This topic is dealt with more in chapter two. Communication has an end as well as a beginning, an audience as well as a speaker. The audience is *part of the message*, and different audiences can transform the meaning of the same message. Robinson writes:

> 'Ring-a-ring-a-roses a pocketful of posies' would convey different messages uttered by a teacher to five-year-old children in a games lesson, by a teacher of linguistics illustrating the achievement of poetic or rhythmic effects, and by a managing director of an industrial company at a board meeting at which the members appeared to be squabbling unnecessarily[15]

Communication channels have other important characteristics. The greatest gulf in the world is between two minds, and sending a precarious leaky cargo of ideas over that gap is a permanent challenge. The channels also have noise, require some redundancy, and benefit from feedback.

In a speech, *noise* is any form of interference with the message which produces extra and distracting information. It can be as simple as the roar of traffic through an open window, it can be as obvious as the intolerable heat, the buzzing of flies, and the desire to sleep that comes over an audience listening to a droning speech after lunch. It can also be as subtle as an unwary reference to an audience's pet hate. Anything which interferes with easy communication has the effect of noise in a communication channel.

Redundancy is also a feature of all human communication. Stylists are often heard to recommend economy, and despise vacuous repetition. But a speaker describing the shape of equipment with his hands as well as his voice is using redundancy helpfully. Saying the same thing in different ways, using different media, or simply saying it twice may be a useful aid to clarity.

Feedback is a vital component of successful speaking. By being sensitive to the audience's reactions, the speaker can modify his message to achieve the best effect. All three of these theoretical components of communication, then, have their counterparts in spoken communication. Too many thoughtless speakers think that their words reflect exactly what they are thinking, and that their hearers interpret those words perfectly. Not so. Nothing is perfect, and language is very imperfect.

Speech came first

Spoken language was the first form of communication between human beings. It came long before written language, and writing is a transcript of speech, not vice versa. This more primitive form of communication still provides the most direct access to other minds. The reason why people prefer to listen to a spoken explanation is that it seems to need less effort to understand than the more formal medium of writing. Yet some speakers try to make speech as close to writing as possible, and destroy its freshness and immediacy. Speaking is the direct route from one mind to another, and is the way we usually choose when we want to ask a question, or give an explanation. Research shows that ideas and information are more easily understood and processed through speech than through writing.

Unless they are pretending to be formal, people usually speak in a style which is more direct, and easier to understand, than the style in which they write; speeeh makes the personal interaction more immediate. One of the reasons is that when speaking, interest and enthusiasm in the listeners are generated by non-verbal, as well as by verbal, signals. The variety and impact of the message are heightened by the presence of another person. Listeners also feel more secure when they can see the person who is giving them new information. Their judgement of the validity of the message, the competence, and the depth of knowledge of the speaker is easier if non-verbal clues, as well as verbal clues, are available. There are many reasons why speaking is the best of the communication channels. It is not always used, largely because people are afraid of their inexperience and inability to speak well. Yet practice and study can provide the skill needed to use this most direct path into the minds of others. It is worth the effort to become an effective speaker.

Finally, let me say that I am very much aware that this is a book of advice, and books of advice all too easily become patronizing and repetitive. They tend to be full of sentences of the form '*do* this', '*don't* do that', '*remember* the other', and '*avoid* something else'. This becomes tedious. I have moderated advice with anecdote, and prohibition with discussion. But I cannot change the basic character of the book. I hope you will take it all in the spirit in which it is intended.

I have also tried to avoid sexist language, particularly in the choice of pronouns. In my university there are people in the forefront of the equal opportunities movement, and we have had many discussions about the use of 'he', or 'he and she', and other problems in non-sexist writing. Our conclusion has been that it is not possible to avoid all sex-specific pronouns without producing clumsiness of style. But in the present stage of the feminist movement, that is better than using the older convention where 'he' was supposed to stand for both men and women. I have tried to mix these approaches; I hope it will demonstrate a recognition of the fact that successful executives, administrators, technologists and speakers are as likely to be women, as men.

In the main, I restricted the references to books which are easily available in paperback. You will find these are the best sources if you want to read more about a topic. There is also a short list of further reading at the end of the book, as well as a detailed bibilography which is intended for specialists and students. Each chapter ends with a one page summary of the key points in the chapter – a sort of crib sheet for people who want reminding, or who are short of time.

Notes to chapter one

1. See: Halle, Morris., Bresnan, Joan, and Miller George, (eds), *Linguistic Theory and Psychological Reality* (MIT Press, 1978).
2. For example, Adler, Ronald B., and Rodman, George, *Understanding Human Communication* (Holt, Rinehart and Winston, 1982).
3. W.P. Robinson, *Language and Social Behaviour* (Penguin, 1974), p.17.
4. George A. Miller, *The Psychology of Communication: Seven Essays* (Basic Books, 1975), p.91.
5. Hans and Michael Eysenck, *Mindwatching* (Michael Joseph, 1981), p.8.
6. Michael Argyle, *The Psychology of Interpersonal Behaviour* (4th Ed., Penguin, 1983), pp.11–12.
7. For a recent introduction, see: Quinn, Virginia Nichols, *Applied Psychology* (McGraw Hill, 1984).
8. Quoted in Donald Bligh, *What's the Use of Lectures?* (Penguin, 1971), p.9.
9. See: Fransella, Fay (ed.), *Personality: Theory, Measurement and Research* (Methuen, 1981).
10. Wicklund, R.A., Objective Self-Awareness, *Advances in Experimental Social Psychology*, Vol. 8 (1975), pp.233–275.
11. For an interesting discussion of confidence, see: Powell, John, *Why Am I Afraid To Tell You Who I Am? Insights On Self-awareness, Personal Growth and Interpersonal Communication* (Fontana, 1975).
12. Ruth Beard, *Teaching and Learning in Higher Education* (Penguin, 1976), p.8.
13. George Miller, *The Psychology of Communication*, (Basic Books) p.125.
14. Claude Shannon and Warren Weaver, *The Mathematical Theory of Communication* (Illinois University Press, 1949), p.34.
15. W.P. Robinson, *Language and Social Behaviour*, (Penguin) p.28.

Further reading

Three recent books on spoken communication skills are:

Robinson, Don, and Ray Power, *Spotlight on Communication. A Skills Based Approach* (Pitman, 1984).

Ross, Raymond S., *Essentials of Speech Communication* (2nd ed., Prentice Hall, 1984).

Verderber, Rudolph F., *Communicate!* (4th ed., Wadsworth, 1984).

SUMMARY SHEET

Chapter one – Communicating

Everyone has to speak sometimes.

Speaking can be learned.

Research evidence helps us to understand.

Sometimes research results are surprising.

Self-awareness is needed to be a good speaker.

Critical advice from a friend is the best quick guide to speaking.

Models of communication show:
— Understanding, not just words, should be the aim
— Select what you say for the audience's benefit
— Noise is anything which interferes with the message
— Redundancy is common in all forms of communication
— Feedback comes from the audience's reactions.

Speech was the first, and is still the most important, way of communicating ideas and information.

Books of advice can be patronizing – hopefully not this one.

Non-sexist language sometimes sounds clumsy, but is more accurate.

Notes on further reading are listed at the back.

2

The audience

Think about the audience

Thinking about the audience is the first stage in preparing to give a successful talk or presentation. They are the recipients of the information; it must be selected and tailored for their needs. They are also the people whose presence will make you nervous when you speak, whose reactions will depress or encourage you, and whose judgement will measure your success or failure.

When you are thinking about this audience, you must remember, too, that they are active, not passive, participants. They are not empty jugs, sitting waiting for you to pour information into their ears. They have attitudes, interests, likes and dislikes of their own. So the speaker has a personnel management role; he or she has to deal with people and not just with facts. He must not only dole out the information, but anticipate difficulties, deal with problems, to smooth the whole process. So what does a speaker need to know about his audience?

Firstly, he or she should be aware that all audiences have some of the qualities of a crowd. An audience is a group of individuals, many of whom the speaker may know personally, yet collected together they acquire a new personality. When individuals are collected in a room, in enforced silence, all facing one other individual, the speaker, they change. For instance, it is obvious to anyone who watches an audience that their emotions, such as laughter, boredom, and enthusiasm, are both stronger and more sustained.

Every group, even a small and decorous collection of familiar colleagues, displays some of the qualities the sociologists call 'crowd phenomenon'. W.J.H. Sprott, in his book *Human Groups*, writes: 'There is general agreement that a person who is a full member of a crowd . . . is likely to behave differently from the way he would behave if he were by himself.'[1] The differences of behaviour can be summed up in two ways. The first is that there is a heightening of emotionality. The man in the presence of danger feels frightened; in the presence of other people experiencing and showing the same emotion, his fear is even stronger. The second way is that people in a group have a reduced sense of responsibility, less critical sense, and weaker self-control.

There has been much research to try to determine why it should be that people in groups behave less responsibly than individuals. Miller and Dollard point out that as we grow up we are rewarded when we act in the same way as other people act, and punished for non-conformity. The result is that we are taught to accept leadership from others. People in crowds often behave in ways which they would consider reprehensible if they were alone. The crowd becomes their 'super-ego'.[2]

There is no doubt, then, that a group of people is different from an individual, or even two or three people. Hopefully, no speaker during his regular work as manager, administrator, or scientist is likely to encounter a lynching-mob. But he should not forget that every group is tinged with the crowd phenomenon. Collections of people must be treated with care.[3]

The care is best expressed by spending time thinking about exactly *who* they are, and *what* they want. Most speakers have a fair idea of what sort of audience they have to face. They know, for instance, if a group is likely to be hostile or welcoming. But many speakers do not think long enough, or clearly enough, about their audience. Cumbersome though it seems, I believe strongly that thinking about the audience should be done on paper. The effort of writing explicit answers will crystallize half perceived ideas. An example of a questionnaire which can be adapted to your own circumstances is shown in Fig. 2.1.

Think about the context

First, you should analyse the occasion. Decide what the purpose of the meeting is. What is the audience expecting to gain from being there? Are they hoping to make a decision, or are they there simply to keep an eye on progress? Is the talk of general interest, or is it to give new information about a specific process? Will the audience use the information immediately, and if so for what purpose? Many presentations are chiefly psychological in aim. The intention of the monthly branch meeting, for instance, is often to make sure that people come together at least once a month. It helps to give them a sense of corporate identity, and to encourage their loyalty and enthusiasm. Such a meeting may be a platform for news about the company, a place to set new sales targets, for giving information about progress in meeting these targets, and for news about colleagues.

---

o **Why are the audience here?**

o **What will they do as a result of the meeting?**

o **Is the meeting one of a series?**

o **If so, what are the main errors and triumphs of previous talks?**

o **Are there precedents?**

o **How many people will there be in the audience?**

o **What interest do they have in the talk?**

Fig. 2.1 Audience analysis

Another type of meeting is the symposium of a learned society. Here the purpose is probably to disseminate information and to encourage other workers. Listeners may pick up ideas which apply to their own work, or they may simply expand their general knowledge. Other groups may consist of a few research managers, one or two people from head office, and the speaker's own immediate boss, who wants a new project explaining. It may need the approval of all the audience if the company is going to be persuaded to spend money on it.

There are as many purposes as there are meetings. It is naive to imagine that the purpose is often a single one. I doubt if many presentations are purely for general interest; or indeed if many of them are to sell one particular idea only. They will also be goodwill exercises for the company or department, career-building opportunities for the speaker, and general back-patting, congratulatory sessions for the group. What people will do as a result of the talk is as diverse as their reasons for being present. Some will go back to their offices and sign cheques or requisitions; some will merely forget the whole thing; some will find that in a conversation days later they have information unexpectedly relevant to what is being discussed. The task of visualizing, quite specifically, why people are there is an important step in understanding the audience. Unless you can write down a statement of what the audience will actually do as a result of hearing your presentation, you have not really clarified the purpose of the meeting.

A next question to ask is whether the meeting is one of a series, or whether it was called to deal with the topic of the moment. Are there precedents for such a meeting? Does management ask for regular presentations on research topics? Are administrative bottlenecks always thrashed out in head-of-department meetings, with the responsible officer addressing the group? The attitudes and expectations of the audience will depend very much on what they are used to. Imagine yourself being asked to give a paper; your own knowledge of the precedents will help you to avoid obvious pitfalls. If your paper is to be given in one of a series of research colloquia, it will help to remember your impression of the other speakers you have heard. The audience will probably view you in the same way.

Perhaps the worst feature of the colloquia you have attended so far has been the blind specialization of some of the speakers. They may have been wrapped up entirely in the fascination of their own techniques. The only bit of the last talk you enjoyed may have been, for instance, a short section on the translation of pure research ideas into commercial reality; the rest was irrelevant and therefore boring. From your own reaction to others, you have a model with which to design your own talk. Clearly, in the situation we are discussing, unless there are many people in the audience working on the same specialization, the speaker should keep discussion of the intricacies to a minimum. But information about the commercial hopes and pressures that fuel the research, and their effect on the direction of the work, could form a major section of the talk.

Let me take another example. Imagine a computer systems analyst, presenting technical (not specifically sales) information on a new product for a potential customer. His branch manager may also be in the audience, so it is a career opportunity as well as an information giving session, and obviously an occasion to impress the expertise and quality of his company's professionalism. But if the presentation is one of a series given by every major computer manufacturer competing for the order, a shrewd guess at the line taken by other speakers will help greatly. To repeat the same claims, and offer the same facilities is useless. What is distinctive must be stressed.

Awareness of precedent is essential for a successful presentation. Most talks fit into a familiar context; they form part of a pattern, and the audience's expectations are formed by this pattern. All communication depends on contrast with its context, and language operates by using the contrast between different sounds to signify meaning. For example, the difference between 'red' and 'led' lies only in the first letter. Orientals find the contrast between these sounds difficult to perceive, usually, and without it meaning is lost. Equally, unless there is a contrast between the communication medium and the context it is received in, no meaning can be transmitted.

A language which consisted of a series of humming tones might work well in the quiet plains of Mongolia where we can imagine it originating. It would be useless in a modern factory filled with machinery. Contrast between elements in a language, and between the language and its context is essential. In the same way, yet another paper read in a droning monotone in a conference filled with monotonous papers will not communicate. It will be ignored and forgotten. In considering the precedents for your presentation build on the contrasts that will make it stand out.

How large an audience?

Next, think about how large the audience will be; try to write down a rough estimate of the numbers. There is a scale of sizes (and types) of audience, from a little group of three or four in a small office, through a seminar of twelve to fifteen in a meeting room, to an audience of forty or fifty (or even hundreds) in a large hall. The formality of the presentation will, of course, vary with the size. It is useless to have a largely written script, full of formal language, for the group of two or three in an office. An informal summary, followed by a discussion will be best for them. Equally, a heart-to-heart chat, with little structure and invitations for questions very early on, will fail in a lecture theatre filled with two hundred experts. One interesting result from sociological research shows that as 'group size increases, member satisfaction decreases.'[4] Speakers should, therefore, be aware of the effect of group size on the audience's satisfaction, as well as on the speaker's nerves. Both the ideas, and the voice which accompanies them, have to be bolder and more forceful in a large group.

Think next about the audience's interest in the talk. It will depend on factors such as their age, their status, and their background, as well as their reasons for being there. Were they compelled to attend? What do they expect to gain from the presentation? Most audiences will have various layers of interest. They may have a primary interest in the subject of your talk, but they will also have a secondary interest in other matters, such as the group you work for, and they may have a passing interest in other areas which you talk about. You will also need to know whether they have power to do things as a result of hearing your talk. What can they do for you, or you for them, which forms a community of interest (in both the involvement, and the curiosity sense of that word)?

Considering these factors will not, of course, guarantee success. Indeed, so complex are human interrelationships that not even a team of sociologists could tease out the full niceties of an audience's attitudes and expectations. But the speaker should not abandon attempts to be rational about his presentation, just because an audience is a complex entity. His analysis of the audience will always be imprecise. But this does not matter, because the audience will come half-way to meet the speaker. From their end of the communicative relationship they will be making the same allowances and adjustments as the speaker is making from his. Few audiences are malicious, and the speaker can count on a reserve of willingness and tolerance from them.

We should also realize that language is a very approximate medium. Even if an exact specification of the audience's attitudes and needs could be written, the encoding of the message can only be approximate. So even a crude analysis is better than none. What is needed is protection from the grosser and more obvious mistakes. To plunge in, without having first thought about the audience, is like navigating over a reef without a map. The speaker may make it, and never know how close his or her hull came to the fangs of rock beneath. But it is just as likely that he or she will end up with a wrecked argument, and the cargo of ideas just so much flotsam washing uselessly around in the stormy minds of the audience! If that happens, the speaker has only him or herself to blame if he or she had not first charted the passage. Thinking first, even though it is approximate, is better than making

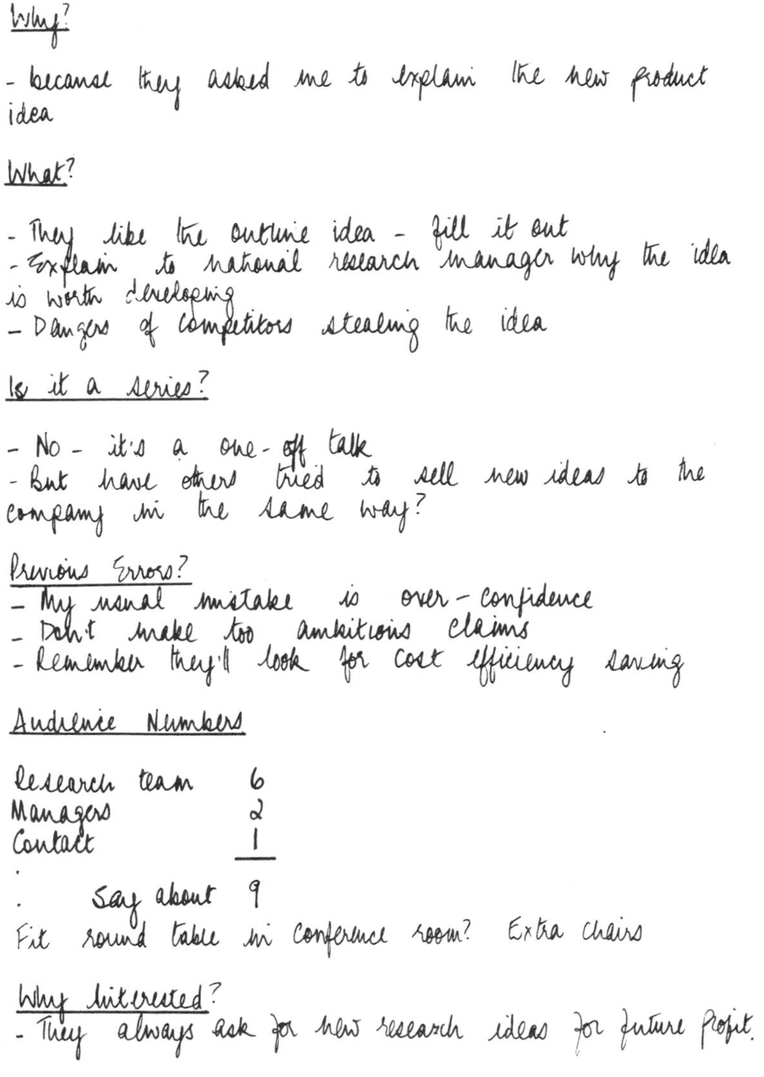

Why?

- because they asked me to explain the new product idea

What?

- They like the outline idea - fill it out
- Explain to national research manager why the idea is worth developing
- Dangers of competitors stealing the idea

Is it a series?

- No - it's a one-off talk
- But have others tried to sell new ideas to the company in the same way?

Previous Errors?
- My usual mistake is over-confidence
- Don't make too ambitious claims
- Remember they'll look for cost efficiency saving

Audience Numbers

Research team	6
Managers	2
Contact	1

. Say about 9
Fit round table in conference room? Extra chairs

Why Interested?
- They always ask for new research ideas for future profit.

Fig. 2.2 Explicit thinking about the audience

mistakes. And thinking is only complete if it is made explicit.

My main advice at this stage, then, is to *write down* the answers to the questions in Fig. 2.1: a typical set of answers for a paper is shown in Fig. 2.2. If you are too lazy to do this, you are almost certainly too lazy to think through the analysis fully. If you are serious about giving a good presentation, take the time to jot down your judgement of the audience as the first page of notes you make towards the presentation.

The structure of groups

I have already suggested that an audience is a very different thing from a collection of individuals – it is a group, and as a speaker, your task is to manage this group, and present your material in a way which will help them. So the more knowledge of group behaviour a speaker has, the better he or she can do their job. I do not propose to write a textbook of group dynamics; there are already plenty of good ones. In any case, the basics are simple enough. But a few pages about groups is a useful preparation for speaking. Thinking about group psychology will help you be more aware of group dynamics, more sensitive to feedback from the group in front of you, and more able to make effective preparations to speak. There is a great deal of research on group behaviour. Even in 1968 a writer noted that there were well over 2,000 different research studies on the topic. Knowledge about groups is now extensive. In many cases it reflects what instinct tells us, but not always, and conscious awareness of group behaviour can only help the speaker.

The first fact about groups we must remember is that they are composed of individuals. As speakers, we too easily imagine that the sea of faces in front of us belong to undifferentiated clones. Stereotyping is used by speakers to misjudge audiences, almost as much as audiences use it to misjudge speakers. Speakers tend to think that an audience from company 'X' will be all whizz kids, or an audience from company 'Y' will be old and cautious. And this enables the speaker to miss the fact that there may, or may not, be some of both these types of people, in each audience.

The mere fact that the audience are all sitting down, facing one way, tends to deprive them of individuality in the eyes of the speaker. This is a curious mistake, since when we are in the reverse position, sitting listening to someone else talking, we feel our individual identity; our own reactions contrast with the group's. Yet when we are speaking ourselves, we tend to think of the audience as some homogeneous, and powerfully distinctive, object. We think of them as having an overpowering common identity. I suppose it is because we feel so conspicuous as the speaker, that we tend to see the audience as about as undifferentiated as a wall. Each brick may be different, but the total effect is massive. Of course, the truth is rather different, for each person in a group is an individual. They all have their own standards and motives, and many also belong to many other different groups.

The first lesson to be learned about groups, then, is that they are collections of heterogeneous individuals. Indeed, far from being repressed, individual roles and individual differences are often enhanced by the crowd effect, the stronger personalities becoming more assertive in response to group pressure. By considering the role of individuals, rather than of the group as a whole, you will be able to recognize an audience's diverse needs. No talk is likely to satisfy everyone in the group, and an outsider will understand if the talk is angled towards the majority of the group. But it is fatal to talk solely to one sector, and appear ignorant of the needs of the rest of the audience. When analysing the audience, assess the variety of different interests it represents, and try to devise a strategy which will speak to all of them.

One good technique is to alternate different kinds of approach, so that no one group has time to lose interest. Thus, for example, an audience including technical, and lay people, can be dealt with by alternating complex technical facts, with a few sentences of simple explanation. A mixed audience of marketing managers, and personnel managers will all enjoy a presentation which alternates between the marketing prospects of each topic, and the way it will affect production. People are surprisingly tolerant, and will listen to several minutes of information that they don't understand, and which doesn't concern them, as long as they know that the talk will come back to their own interests. By providing a mixture, a speaker can cater for a wide range of interests in his audience. But of course he can't use this technique, or any other technique, if he hasn't bothered to work out who his audience is, and what their interests are.

While considering the audience's individuality, it is important to consider their relationships to each other, as well as their relationships to you. There may be both administrators, and researchers in the audience. Their ages may be both younger and older than yours. They may be more hostile to the departmental manager, sitting silently in the front row, than to you. There is nothing sinister in acquiring this sort of knowledge and skill in your understanding of groups. It does not represent 'some dark power that enables them to manipulate people more easily.'[5] Knowledge about groups is a wise part of the effective speaker's armoury. It can be used to avoid making obvious mistakes, antagonizing people, and failing to explain the information. It can smooth an interaction, and help to make the talk a more satisfactory experience for every member of the audience.

Knowledge about groups, then, is also knowledge about individuals. You should learn to identify the assertive (and the 'invisible') members of a group. Talk to the quiet and self-effacing ones *more* than to the alert, responsive and pushy ones. Remember that the status of an individual influences a group's norms powerfully. If there is a very senior manager, a highly respected scientist, a powerful union leader, or an ambitious local politician in the audience, the rest of the group will be aware of him or her. They will also modify their behaviour so they seem typical of the kind of group he expects. Your role as speaker is to be aware of the restraining influence of high status members. The aim must be to counter the subconscious tension caused by the presence of the powerful person, and to help the group to relax. You can do this best by treating the special person as neutrally as possible. Don't show him or her special attention, but equally don't pointedly ignore them. Try to show by your distribution of attention that you regard the person as a normal member of the group, for the duration of the talk. By treating a powerful person without embarrassment and without pointed attentiveness you will help the whole group to feel at ease, and listen to the presentation.

Find the leader

You should try to identify the devious members of the group, and be aware who is functioning as group leader (he or she may be the boss, the manager, or simply the funniest, or most loud mouthed person present). It is worth noting that there will not necessarily be only one dominant figure in the group. Sociologists discovered, when they started studying groups, that there were usually 'two leadership functions: the task specialist and the socio-emotional leader.'[6] There may be one member of the group who is the acknowledged expert, whose approval or irritation at what you say will be echoed by the whole group. A separate person may be the social leader, and if he or she laughs the whole group will laugh. Of course, you must not pander to these leaders; but they are useful indicators of the group's reactions to the presentation. It is usually easy to identify the people who fill these roles. They are often the more attentive members of the audience, apparently taking their function as group leader responsibly. Other members of the group can often be seen glancing at them. The leaders, on the other hand, will rarely look at other members of the group.

Be aware, too, of the challengers for the leadership, who may disrupt the presentation by fighting their own battles for attention, and trying to score group approval. There will also be scapegoats, whom the group may want to assist you in discomforting. Be careful. The group will not necessarily admire you for pandering to their own prejudices. You will also need to learn to pinpoint both the narks and the nodders, and to recognise the tactics of opposition. Notice that some members of the group will be visibly secure in the situation (relaxed posture, crossed legs, and forward positions in the room). Others will be visibly insecure (tense posture, face partially covered by hands, and back-to-the-wall seating position). Try to encourage security, since it will aid the audience's concentration. Try to unite the group, by acting on the individuals in the group to even out their differences. It is much easier to talk to a homogeneous group, and you are more likely to be able to galvanize their interest and enthusiasm.

Another factor is the way in which in larger groups, size brings pressures to bear on the individuals in the audience as well as on the speaker. The effect of size is to make individual listeners more nervous about speaking up, during question or discussion times. But 'unhappily it is not always the right people who are silenced . . . the weight of the group may act like an ox sitting on the wrong tongues. The bore, alas, is seldom unnerved; that, indeed, is why he is a bore.'[7] The speaker's role in such a large group is to function like a chairperson, evening out the pressures. He or she can do this, even during a presentation where the audience is silent. He can facilitate attention from the inattentive, and discourage the disrupters and bores.

The best way to do this is to look at the inattentive, smile, in passing, at the apparently reticent and withdrawn, and pay little attention to those who are nervous, or too highly charged. The enthusiastic nodder should receive only a fleeting glance, but your gaze should repeatedly stop at the bored or withdrawn person. The speaker has to perform a 'delicate balancing act'.[8] He or she must avoid outraging the group leaders, and the members of high status.

But he must also make sure that the less conspicuous group members feel that they are important, and that their interest and attention is valued. The length of dwell of carefully distributed eye-contact is the main skill a speaker employs to unite and balance his group.

Audience receptivity

Half the inexperienced speakers in the world think that the audience will sit listening, uncritically absorbing every word they say, like a huge sponge. The other half have never even thought about it. But does an audience uniformly listen to everything the speaker says? Do they really hear every word? The sad fact is that they do not. Attention is not the simple, conscious activity we would like to think. Everyone's mind wanders constantly.

The novels of James Joyce and Virginia Woolf contain a surprisingly accurate portrayal of the inner mental life of ordinary people. Bloom, or Mrs. Dalloway, do not pay attention to one thing at once; their minds are a stream of ever varying ideas and impressions, where one thing after another floats to the surface with little apparent connection. While your audience's minds may not be as interesting as Joyce or Woolf's heroes and heroines, they will certainly be no less unreliable! Attention is simply not the steady beacon we would like to think. An audience is more like the lines of yellow warning lights on a motorway; their attention flashes on and off randomly at unpredictable intervals. Imagine unsynchronized flashes of light mingling with a background of darkness, as attention switches on and off. This may sound a harsh image, but it contains a large truth. Speakers cannot assume that everyone is listening all the time.

One of the reasons why people do not listen to every word that a speaker utters is that their minds can go much faster than the speaker's voice. It is an obvious point, and it is confirmed by the fact that everyone's reading speed is much higher than their talking speed. While sitting listening to a presentation, the audience's minds have spare capacity, which they will fill by watching what else is going on, and observing the non-verbal components of the speaker's message. They can also use this spare capacity either to pursue private thoughts, or to ponder the remoter implications of the topic. They are certainly not sitting with vacant minds, waiting for the next word.

This spare capacity is prey to being pre-empted by day-dreaming. And once a mini day dream has started, it will be more interesting that the next few words of the talk. The listener will pursue his day dream for a few seconds, before he switches his attention back to the speaker. There has been some interesting research done on attention, and how it is controlled. American neurologists argue that concentration is the result of suppressing countless nerve events which spontaneously trigger trains of thought all over the brain. Untrained mental activity is very largely random.

Training is a process of reducing (not increasing) the brain's activity, clearing the mind so that a single chain of thought can be sustained. The longer this clearing process continues, the more the pressure builds up from other concerns wanting conscious attention. The successful speaker will recognize that this pressure will break through continually, whatever he or she does. The audience will indulge, involuntarily, in 'micro-sleeps' which are momentary rests, gaps in attention, when their minds plunge into a sort of dreamy semi-consciousness, before emerging refreshed to pay renewed attention. We should not blame the audience for their fickle behaviour and paucity of interest. Rather, we should be realistic, and try to be aware of who is listening and who is not.

The listeners, then, are not just passive vessels, sitting with ever open minds. They are, like everybody, full of their own concerns and interests, and prone to drift off for fantasies, day dreams, or private thoughts. Some speakers regard the idiosyncracies of the human receiver as a nuisance. George Miller remarks:

> I have the impression that some communication theorists regard the human link in communication systems in much the same way as they regard random noise. Both are unfortunate disturbances in an otherwise well-behaved system and both should be reduced until they do as little harm as possible.[9]

But such an attitude is clearly nonsense. It is not wilfulness on the part of the audience which makes them imperfect receivers. It is the operation of a natural mechanism. The speaker's job is to recognize, and plan for this fact. You cannot change it.

Attention controlling

There are a number of factors which affect attention, and can be used to control it. These are variety, the length of time concentration is needed, the time of day, and the amount of arousal and motivation the speaker can communicate to the listeners. Motivation, in turn, is affected by the audience's sense of security, how much their natural responses are repressed, and how much enthusiasm the speaker shows. Let me deal with each of these factors in turn, to give you some ideas about how to control the audience's attention.

The first important fact to grasp is that attention is very little under voluntary control. The audience cannot *make* themselves listen – they must be interested. Psychologists tell us that attention is controlled by a deep part of the brain which operates subconsciously. Attention is automatically switched-off by repetitive stimuli:

> If, for example, you are in a room with a clock that is ticking quietly you will quickly habituate to the sound so that after a short while you will no longer hear it. But the sound is still being continually monitored by the brain, and if the clock were suddenly to stop, or to change speed or volume, you would immediately notice it.[10]

A good speaker knows this, and arouses questing interest in the audience by providing a continual variety of stimuli. He or she keeps them alert by varying the input of ideas, and gives them continual change in the pauses, speed and volume of his or her voice. A speaker should not blame the audience if they nod-off, or their attention wanders. They are responding to simple physiological mechanisms; the fault is the speaker's for not being aware of these facts, and not negotiating around them in the techniques of his presentation.

The second factor which affects the audience's attention is the sheer length of time they are expected to listen. In the first few minutes of a presentation everyone is listening. As time goes on the pressures from other thoughts gradually increases. It depends to some extent on the level of training and discipline in the listeners, but the speaker should always be aware of the way the demands he is making on the audience increase with time. One piece of research by MacManaway reported that 84% of students said that twenty to thirty minutes was the maximum length of time they could listen to a lecture without wandering.[11] This means that the speaker should provide pauses, and increasing variety, as the talk gets longer. The vital points are best made early in the talk, when you can count on more attention. Gradually working up to the important point through an hour long maze of details, may gain nothing. By the time you are ready to triumphantly announce your answer, most of the audience will be thinking of something else.

If we are to be effective communicators we should understand, not blame, the audience's natural characteristics. There is no point in battling valiantly against the natural effect of growing inattention.

The solution is to provide breaks, variety and interesting changes throughout the talk. Otherwise the discomfort of sitting still will break through as interest flags; this is signalled by a rash of scratching and shifting about in the chairs. Very high levels of discomfort are produced by forced attention to uninteresting material. Indeed, some members of the audience may develop gestures of extreme misery, such as dropping their heads into their hands, and even groaning quietly to themselves! A considerate speaker (considerate, that is, for his own success as well as for the audience's comfort) will alleviate the misery. Variety, both in the material and in the presentation, is the best method. But variety can only be used at the right time if the speaker first learns to be sensitive to the audience's mood. The sheer length of time listeners must pay attention is an important factor in the equation that measures their ability to listen.

How tired are they?

The third factor which affects attention is the simple matter of the time of day when the talk is being presented. The speaker must think in advance about the effect of time of day on his audience. It is not just the personal experience of the lazy, but a psychological fact that people's intellectual sharpness varies during the day, in response to internal biological rhythms. Both body temperature and hormone levels change in a rhythmic cycle. Most people are at their best in the morning, but some people do not reach their peak until midday, or even during the afternoon. Most people have a low in body temperature and hormone level during the mid afternoon, and are therefore likely to be sleepy, and find it difficult to concentrate.

The day of the week will also affect how easy it is for the audience to concentrate. Almost everyone is better on a Tuesday than on a Friday. Even the disciplined hard workers are feeling the effects of tiredness by the end of the week. These effects are very real, and should not be dismissed as softness. Research showed that a prolonged period of monotony interferes "with the ability to make decisions at a fairly high cortical level."[12] Monotony interferes just as badly with the ability to listen intelligently. The speaker must provide greater variety and stimulation for an afternoon presentation, or one at the fag-end of the week. On Tuesday morning he can afford to rely on his audience's concentration more.

The fourth of the factors which affect attention is the psychological state of the audience while they are listening. The state of mind of listeners can be separated into their arousal, and their motivation. The speaker must be continually aware of the twin factors. Arousal is a technical term in psychology for the level of alertness, the biochemical tone and readiness. It affects the whole performance of the brain, increasing the transmission rates in the neurones, making the person more alert and more receptive. Motivation helps arousal. A highly motivated person is more prepared to understand and remember.

Levels of arousal have been compared to an inverted 'U' curve, where performance levels rise as arousal increases, but dissolve in chaos as arousal gets too high. Most people spend most of their life in the restricted performance, low arousal part of the curve. Speakers spend much of their time in the hyper-active, anxious and over-aroused part of the curve. The art is to get both speaker and audience on the peak performance part of the curve. For speakers, as I suggest in chapter six, this usually means quietening themselves down. For listeners, it usually means awaking their interest, arousing them, by energy and variety.

I am going to suggest four tactics for raising arousal and motivation here, but most of this book is about how the speaker's performance can be tailored to increase the interest of the audience. And interest can be equated to arousal and motivation. These four tactics are sharing interest in the listeners' problems, giving a sense of security to the listeners, recognizing that enforced silence represses levels of arousal, and finally communicating your own enthusiasm.

Firstly, then, share your listeners' interests. You will arouse interest and motivation if you make it clear from the outset that you have considered *their* needs. Thus a researcher might start a paper by mentioning the way his work provides promising analogies to the audience's own work; a manager may open by mentioning the common need for the company's success; a lecturer might mention examinations in his introduction. It is possible to effect the arousal of listeners by talking about things which closely concern them. In this way the speaker can build on the motivation they already possess. One way of doing this is to present trailers, in the cinema tradition, for the points which are to come, brief extracts of the information and ideas you are going to present. The best way is to spend time showing how your ideas relate to the problems, and interests, of your listeners, before launching into your own interests.

The second tactic which affects the acquisition of information, is the degree of security the audience feel in their speaker. All education-alists know that in learning, emotional stability is almost as important as intelligence. A calm and emotionally secure person will think more clearly than a highly intelligent, but emotionally disturbed person. The speaker should consciously try to calm and reassure the audience. Provide a secure atmosphere, and clearly defined physical constraints, such as the use of space (see Chapter Ten), and good timing (see Chapter Seven), to enable them to concentrate. If the audience feel at ease, and if they feel confident that the speaker knows his job as well as his subject, and will stop on time, they will find it refreshingly easy to listen to the presentation. If they are cramped, suspicious, anxious, and unsettled, they will day-dream more, and listen less. Providing the right emotional conditions is an important factor in effective speaking.

The third topic which we should consider is that sitting still and listening affects the natural levels of arousal. One of the disadvantages of the spoken presentation is that it stifles two major needs of the audience, those for self-expression, and those for social interaction. They are expected to be silent and listen, and their ability to interact with others is savagely curtailed. The audience's natural arousal is reduced by these repressions, and extra stimulation must be provided to compensate for this. Careful consideration of the way the audience's receptivity to the message is affected by natural psychologicsl mechanisms will be rewarded by an attentive audience.

Finally, there is much evidence that the speaker's own com-municated attitude and psychological state will affect the audience's reception of his message.

> Mastin instructed lecturers to teach one topic with an 'indifferent' attitude and another the following week 'enthusiastically' . . . Nine-teen out of twenty classes did better on multiple-choice tests after the 'enthusiastic' lesson . . . In a similar experiment Coats and Smidchens found that 36% of the variance in tests of audience recall were attributable to the dynamism' of the speaker.[13]

The conclusion from this research is that an enthusiastic and energetic presentation is more effective than a dull and soporific one. If you are enthusiastic yourself about the subject, it will be reflected in the audience's echoing enthusiasm.

There are a variety of ways, then, in which we can vary the levels of receptivity in an audience. Whereas no one technique for coping with this is a panacea, a good understanding of the natural operation of the listener's mind is an enormous help in adjusting the presentation so it is as effective as possible. Of course, none of this is possible if you have not thought about the audience. Nor is it possible unless you are sensitive to the way your presentation is affecting the audience. Before dealing with feedback from the audience, though, let me explore one more component of the delicate equation between speaker and audience, the way the speaker feels about his listeners, and the way they are likely to feel about him.

The relationship between the speaker and the group

Any communication establishes a relationship, in which both receiver and sender of the message are working out their impressions of each other. When speaking, this relationship is unusual, because the receivers are many, but the sender only one. If the relationship goes wrong, it will make effective communication impossible, both because the audience will be suspicious and uncomfortable, and because the speaker will be unnerved. So the relationship between speaker and group is an important component in the message; it is worth spending time exploring it.

I have already discussed the way in which grouping people can have a strange affect on them. A group of listeners is not, however, interacting solely with itself. It has an outsider (in group terms) to contend with – the speaker! A group of silent people, with a speaker (who is the only one the rules of the situation allow to move about or to talk) inevitably creates perceived dominance in the speaker. There are always strains in the psychological and social balance of a group of people. This is especially so if one person is obeying a different set of rules from other people in the same space.

I once bought cheap Covent Garden opera tickets on impulse from a bedraggled ticket-tout, half-an-hour after curtain up. We were wearing jeans and T-shirts (since we had meant to spend the evening at the cinema). We didn't know until we were inside that it was a Gala night and the rest of the audience, including the Queen Mother, were in evening dress. I can still remember our reaction. We thought then that it was irresistably funny, and we still do. But the one thing we were not was comfortable. The group pressure to conform was painfully obvious. In just the same way, a seated and silent group facing one way transmits anonymous social pressures to an individual facing the other way.

Groups are difficult for an individual to relate to, then. One good friend is blessed companionship. Two or three good friends are fun. A party filled with friends can be a happy forum for our best jokes, and most effective imitations. Even one to two strangers are manageable, since the familiar and well-practised rules of good manners mean that we are likely to negotiate the interaction without difficulty. But introduce one stranger into a group of friends and the atmosphere changes. Introduce many strangers, and our instinct is to withdraw, or at least to be silent and effacing until we have sized up the group carefully. The presence of many strangers for whom we are expected to be an instantaneous group-leader inevitably produces psychological tensions. It would be foolish to pretend that these tensions do not exist, either in the speaker or in the audience. To try to disguise them from ourselves as we speak merely leaves uncharted the reefs we sail over. Much better to recognize the tensions, and plan our strategy to avoid them.

The speaker feels exposed when facing an audience, and it is better to understand the reasons for this, than to try to ignore it. The relationship between the self and others has been much studied by psychologists. We are often, as individuals, not aware of how much our own self-image depends on others. Margaret Mead wrote that, 'The self, as that which can be an object to itself, is essentially a social structure, and it arises in social experience.'[14] We depend on the attitudes of others for our very identity, and we cannot pretend to ignore their valuations. These are facts we should not blink, or be ashamed of. We are all the same, and no speaker, novice or experienced, can escape concern over others' views of them:

> We not only search the eyes of our fellow members whispering: 'How am I doing?', we also look at ourselves and say: 'What do they think of you? Do you stand high or low?'[15]

But we must not allow our fear to override our common sense. Understanding of the basic rules of this interaction will enable the speaker to negotiate it successfully. Most people manipulate these rules in one-to-one situations with a secure, acquired competence. We know that if we smile and look interested we can expect pleasure and interest in our company in return.

Think about your relationship to the grouping represented by the audience. Most obviously, ask if you are outside, or inside, the group. The attitudes of the audience must also be considered. Are their attitudes likely to be favourable to you, the group or organization you work for, and the ideas you are presenting, or will they be hostile or indifferent? What relationship do the audience feel towards you? Do you have the same status, or are you junior, or senior, to them? Are they all members of a customer company, and you the sole representative of your own? Are they all accountants and supervisors, and you the only researcher? Are they all shop-floor workers, and you the only representative of management? Are they all older people, and you the only junior? Or are they all teenagers, and you the only older figure?

Such questions may merely underline obvious features of the situation; but by facing them explicitly we acknowledge them, and can thereby begin to counter their effects. To avoid discussing them with yourself is often to allow anxiety to build up without relief.[16] Try to write down the basic dynamics of the situation, so that at least you know that you can face up to it. The analysis is never simple; in most real situations the audience will not be a homogeneous group. You will share some groupings with them, and be outside other groupings. Figure 2.3 is an example of such notes on the audience.

Such doodling is a great help. It calms the nerves, by doing something about a problem, even if that something is only drawing a picture of it. The activity of drawing releases the dammed-up flow of

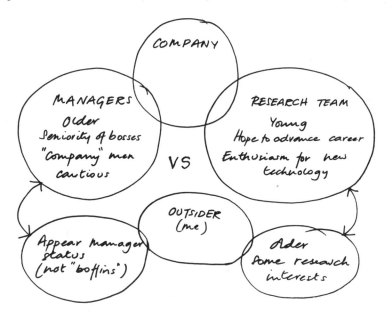

Fig. 2.3 Relationships between speaker and audience

thought and ideas. As we puzzle over the shapes to draw, and how to interlink them, the mental covers which have been clamped over our subconscious fears are released, and useful ideas can rise into consciousness. Do not, of course, expect the puzzle to come out. You will not be able to produce a neat and symmetrical diagram; but you will have clarified your own feelings about the audience, and its attitudes to you.

Stereotypes

Another set of questions speakers must ask themselves is, will they view me as a stereotype? How will they interpret my superficial attributes? Which prejudices will I arouse in them? If you are management and they are union representatives, will their suspicions make them perceive what you intend to be concern for the common good as a trick? If you are a research scientist, will the line-management see you as only a technician? If so, a white lab coat is the wrong costume. If you are a professor, and they are students, will they interpret your friendliness as the pathetic attempts of age to bribe youth?

Speakers are often ignorant victims of stereotyping, so it is worth examining the research findings of psychology to try to understand the phenomenon. The speaker is usually the stranger in the group, yet he is fully exposed to the group's gaze and assessment. It is inevitable that the audience sees him or her as a member of a familiar group, about whom they have fixed ideas and prejudices, in other words a stereotype. One experimental study of stereotyping discovered that a simple conventional stereotype was relaxed as the type-cast person was better known:

> The stereotype of bespectacled people being more intelligent is readily susceptible to elimination if further information is provided. Wearing spectacles increases a target person's rated IQ by about 2 points when seen briefly, but has no such effect when he is seen talking for 5 min.[17]

Better acquaintance usually results in improved judgement of the real person. But the effects of immediate appearance, and the stereotyping it evokes, are very real. Michael Argyle warns that:

> The constructs an individual uses may be extremely weird and private. Whole groups have their constructs, like 'saved' and 'not-

saved'. Some people use very simple category systems, with only one or two dimensions, such as 'nice – nasty', in the Army – not in the Army'.[18]

Thus we can expect some members of some audiences to be very rigid in their perception of stereotypes. It may be possible by continued exposure to break down their prejudices, but it is rarely worth the wasted time for the speaker if the stereotype can be sidestepped instead. Far better to be aware of what these stereotypes are, and modify your performance to produce the impression you want. Stereotyping is, of course, at the root of prejudice, and speakers will often meet prejudice. If all you know about someone you are going to meet is that they are at Cambridge, teaching Theology, and have never married, you will already have a stereotype of the person you expect to see. Katz and Braly did a famous study in 1933, which was repeated in 1951 and 1967, on the stereotypes of Princeton students: 84% thought negroes were superstitious, 79% thought Jews were shrewd, and 78% thought Germans were scientifically minded.[19]

These prejudiced stereotypes apply also to accents. Sociolinguists who study the social markers in language, have done much work on class and regional accents, and much of it confirms, in weighty detail, the prejudices we are all aware of.[20] Strongman and Woosley, for example, conclude their study of the way regional accents affect people's judgement of strangers in this way:

> Both groups judged the Yorkshire speakers to be more honest and reliable than the London speakers and the London speakers to be more self-confident than the Yorkshire. Northerners judged the Yorkshire speakers to be more industrious and southerners judged them to be more serious than the London speakers. Northerners also judged the Yorkshire speakers to be more generous, good-natured and kind-hearted than the London speakers, whom they rated as slightly more mean, irritable and hard.[21]

For people who suffer unfair judgement from such prejudice, the persistent force of stereotypes is exasperating. But we also know that confronting the stereotype with contradictory evidence reduces and finally demolishes it. The stereotype moves towards the reality; people get to know what you are, and forget what you seem. But by recognizing the stereotypes the speaker can avoid inviting the audience to reinforce those stereotypes by behaving in a way which they see as typical. As I remarked when discussing audience attention, there is little point in a speaker attempting to blame the audience for its natural vulnerability to well-known psychological facts. Far better to recognize what these facts are, and negotiate skilfully round them. The speaker is not a missionary; he has a different job of work to do. He must communicate information, and battling with prejudice will get in the way of that job.

In order to make explicit the hidden fears about stereotypes, write down what you expect your audience to see in a speaker with your characteristics. We all know, better than we admit, what our characteristics are. We can soon guess, if we are prepared to look steadily at the facts, what the prejudices will be. Identify the stereotype, analyse it and plan to counteract it. The talk will be much more successful.

Finally, recognize that speaking is a heightened version of familiar day-to-day interactions. In all our relations with others we are performing. In most cases the performance is so habitual as to be indistinguishable, for us or others, from our real selves. But when speaking, unless we are aware that we are performing, we may fail. A flat or indifferent performance when many others are watching, receives little sympathy. The speaker is like an actor on a stage. His professionalism must include a certain largeness, an ebullience, a confidence and grasp of himself and his material, which can hold the audience's interest. The speaker must always remember that he or she is stage managing an impression. The performance must be maintained, and lapses or gaps in the performance carefully covered. Erving Goffman gives an entertaining, and worrying list of the ways in which this performance can be shattered:

> Three rough groupings of these events may be mentioned. First a performer may accidently convey incapacity, impropriety, or disrespect by momentarily losing muscular control of himself. He may trip, stumble, fall; he may belch, yawn, make a slip of the tongue, scratch himself, or be flatulent; he may accidentally impinge upon the body of another participant. Secondly the performer may act in such a way as to give the impression that he is too much or too little concerned with the interaction. He may stutter, forget his lines, appear nervous, or guilty, or self-conscious; he may give way to inappropriate outbursts of laughter, anger, or other kinds of affect which momentarily incapacitate him as an interactant; he may show too much serious involvement and interest, or too little. Thirdly, the performer may allow his presentation to suffer from

inadequate dramaturgical direction. The setting may not have been put in order, or may have become readied for the wrong performance, and may have become deranged during the performance; unforeseen contingencies may cause improper timing of the performer's arrival or departure or may cause embarrassing lulls to occur during the interaction.[22]

The list is not meant to inspire fear, for such disasters are common, and no speaker is immune from the occasional lapse. But he or she does have to be careful. The speaker, whether he likes it or not, is thrust into the role of the performer. His qualities and successes will be judged by the rules of performance, and the cruelties of stereotypes. Resenting this has no future. Understanding the reasons for people's reactions is a better path; after all if we are honest, we all judge others in the same way that they judge us.

Feedback

The problems of interacting with an audience can never be solved entirely in advance. Preparation is always needed, of course, and a speaker who has not thought in advance about the composition, interests, and attitudes of the audience is not likely to achieve success by sheer chance. His talk is already doomed to some misjudgement or other. But even when careful preparation has been made, there is one further element in the equation. Watchful attention to the way the audience is reacting during the talk itself will give you a chance to correct mistakes, and fine tune your judgements about what does and does not need saying.

The feedback the speaker gets from his audience while he is talking is the last component of effective speaking. Many naive speakers ignore the signals from the audience completely. They may not even be looking at the audience, but be gazing nervously at their feet, or staring airily at the ceiling. By failing to study the audience such speakers miss the vital non-verbal signals that should guide the shaping of the talk. I suggest in the next chapter that the range of examples you use, and the speed at which you run through new or difficult ideas, should be controlled by the way the audience responds.

How do you tell how the audience is responding? You look at them. Only by knowing how the audience is taking what you are saying, can you modify it to fit in better. You need to know whether your listeners understand, believe you, are surprised or bored, agree or disagree, are pleased or annoyed. You find this out by studying their faces, as you do in normal conversations, watching specially their eyebrows and mouths.

Throughout the talk, then, the speaker must look at the audience, trying to gauge their reactions, and observing and using the feedback their expressions and movements provide. In fact, every speaker knows underneath what the audience is thinking and feeling, even if he won't admit it to himself. If the audience is getting bored, there will be an increase in shuffling and scraping noises; there may be a rash of yawns, their eyes will be lifeless, their expressions fixed, and the mouths hard and unsmiling. It is usually quite obvious when the talk has become tedious to listen to, yet boring speakers carefully ignore such signals. Perhaps it is a form of self-defence; they are safely wrapped up in their own world. Yet such defence is illusory; ignoring the audience's boredom will not make it go away. The only recipe for success is to alter your tactics according to the audience's reactions.

An interested audience also makes its state of mind clear. There will be a watchful silence in the room. The listeners' eyes will be looking at the speaker, and will be mobile and alert. There will be slight wrinkles of smiles around the eyes and the mouths. There may also be slight nodding of the heads, and gestures such as leaning forwards to hear better. Ready laughter will greet jokes, and small responsive noises (best represented as 'hum', and 'ha', and 'tut') will be discernable. The way to achieve this kind of response is to modify what you say, by watching the feedback they provide.

If a speaker ignores audience feedback, he or she is not really interacting with people, he is merely declaiming. If the speaker is showing no interest in the audience, how can he expect the audience to show interest in him or her? The best tactic to produce this desirable state of alert enjoyment is to be interested in your audience, and in getting your subject across to them, not just in getting through the time. Try to show that you understand their problems. Don't be insulting or patronizing, by ignoring their share of the communication. Respond to the feedback they provide, and be alert to the signals they give off. Then the subconscious signals every speaker communicates by gesture and expression will be right.

If you find it difficult to think of the audience when speaking, it is useful to begin the process of sensitizing yourself to audience reactions when you are listening to someone else's presentation. The best way to learn to be a good speaker is to learn to be a good listener. Whenever you are listening to a talk, take an interest in audiences and their reactions. This will teach you to be observant and considerate (i.e., successful) when speaking. When bored with a talk you are listening to, watch the audience. They smile when the speaker smiles, shuffle and look around when he loses animation, and day-dream when his voice lacks variety and becomes monotonous. Learn to identify, locate and recognize the signs of an audience's reactions. Then as a speaker you will be more sensitive to them.

One final point on feedback from the audience. When you are speaking, you must pay attention to the audience as individuals. But you must not single out one person only, and then ignore the rest; you will embarrass the person you choose, since you are talking to him or her in the watchful presence of everyone else. The audience will also resent the implied favouritism and cliquishness since you will almost certainly have chosen a friend. He or she undoubtedly knows the gist of your message already, so you should be talking to others. A speaker must communicate with all the group, not with just one. This even handed distribution of attention is shown by looking around, not frantically, but steadily and evenly. If you look at everyone, you will also be receiving the feedback from everyone. By being aware of people's reactions, and looking at everyone, you are forming part of a communicating relationship with the whole group. This sense of interacting with the audience, not just of confronting them with words, is the basis of effective speaking.

Feeling with them

Having identified your audience, and thought through its likely attitudes, try to feel understanding for them. Empathy between audience and speaker is a vital ingredient of successful spoken communication. Avoid feelings of hostility, distaste or anger towards your audience's known attitudes. Look for their justification, and for the humanity in their beliefs, and work towards insight. Few people are aware how important their own behaviour is in their dealings with others until they try to give a spoken presentation. Speakers who are aggressive, glum-looking, or nervous, will repeatedly encounter audiences which are resentful, dull, or embarrassed. The cure lies in themselves.

Smiling openness will usually meet cheerful interest and sympathy in the audience. But, on the other hand, the audience will (quite rightly) dislike a speaker who is overconfident and brash. We all know that loud, pushy behaviour is a form of compensation, but we still find it difficult to sympathize with. The opposite tactic will fail too. Nietzsche said that 'sympathy is the last weapon of the weak,' and the speaker who launches into self-deprecation will not even achieve the weapon of sympathy. For self-deprecation is unattractive and ineffective. The audience will rapidly agree that you're not much of a speaker if you repeatedly tell them so. Remember that modesty, self-effacement, and forgivingness are not always what they seem. A *Playboy* interview with Bob Dylan makes the point amusingly:

Playboy: How do you get your kicks these days?

Dylan: I hire people to look into my eyes, and then I have them kick me.

Playboy: And that's the way you get your kicks?

Dylan: No. Then I forgive them, that's where my kicks come in.[23]

This advice about avoiding spurious modesty does not preclude openness. If it is the first presentation you have ever given, say so simply and honestly, without blushing or asking for special favours. To pretend to be experienced and confident when you are in fact inexperienced and terrified can be as fatal with an audience as with a new girlfriend. Ask yourself whether what you want from the audience is pity, or comprehension. Concentrate on the factual content of your talk to help calm your nerves, and reassure them. Three reflections will help: firstly, most audiences are extremely tolerant of minor, but well-meaning, lapses of skill. Secondly, general standards of speaking are low, and thus standards of expectation are also low. And, thirdly, audiences will always help beginners; they are more embarrassed than the speaker is, if the talk breaks down.

Finally, if I were asked which was the main advice I would give a novice speaker, I would chose these three:

1. Trust and like the audience, do not fear and confront them;
2. Look at them;
3. Smile.

The second and third of these are, of course, the ways in which the first is expressed. Even if you do not feel trust and liking, if you look at them and smile, you will probably be credited with having that feeling. You may also gradually acquire the genuine feeling. Eye contact is discussed in chapter nine. But second to failing to look at the audience, the commonest fault (which has ruined talk after talk I have watched) is failing to smile. By that I do not mean that the speaker should adopt a constant grin. But he should try to look happy. Enthusiasm helps. Nineteen out of twenty classes did better after an enthusiastic presentation than after an indifferent one on tests of learning.[24] One reason for this may be that the audience feels that an enthusiastic speaker is rewarding them for their interest. Anything which makes the recipients feel that an interest is being taken in them personally, is a reward. A smile is a reward. Silence, a sigh, or a disapproving look is taken as punishment.

Nothing dooms a talk so quickly as a grim look. The audience interpret it as, at best, severe indigestion, and at worst misery and hatred. Smiling is never easy when you do not feel like it. When you are tense and nervous it is too easy to look like a crocodile! Yet the effect is never as bad as it feels. The subtleties of muscular tensions (which you can feel all too clearly) are invisible from the back of the room. And in any case a nervous, but genuine, smile can be disarming. The good intentions are plain, even if the result is not worthy of Hollywood. If you think about presentations you have enjoyed, many of them will have been given by smiling people. Watch carefully at the next good talk you attend. Good speakers smile, and when they do a quick glance at the audience will usually show many in the audience unconsciously smiling back. Smiling is vital. Many dull talks fail because the speaker's apparent misery makes them a trial for the audience, too.

Notes to chapter two

1. W.J.H. Sprott, *Human Groups* (Penguin, 1967), p.160–161.
2. Miller, N.E., and Dollard, J., *Social Learning and Imitation*, (Routledge and Kegan Paul, 1945).
3. See Douglas, Tom, *Groups: Understanding People Gathered Together* (Tavistock, 1983).
4. See Hare, A.P., Interaction and consensus in different sized groups, *American Sociological Review*, Vol.17 (1952), pp.61–267; Slater, P.E., Contrasting correlates of group size, *Sociometry*, Vol.21 (1958), pp.129–139.
5. M.L.J. Abercrombie, *The Anatomy of Judgement* (Penguin, 1979), p.97.
6. W.P. Robinson, *Language and Social Behaviour* (Penguin, 1974), p.130.
7. W.J.H. Sprott, *Human Groups*, (Penguin, 1967), p.117.
8. Erwin Bettinghaus, *Persuasive Communication* (Holt, Rinehart and Winston, 1980), p.212.

9. George A. Miller, *The Psychology of Communication: Seven Essays* (Basic Books, 1975), p.51.

10. Peter Russell, *The Brain Book* (Routledge and Kegal Paul, 1980), p.43.

11. L A. MacManaway, Teaching Methods in Higher Education – Innovation and Research, *Universities Quarterly*, July 1968, pp.327–336.

12. James Macworth, *Vigilance and Habituation* (Penguin, 1970).

13. Donald Bligh, *What's the Use of Lectures?* (Penguin, 1971), p.80.

14. Margaret Mead, *Coming of Age in Samoa* (Penguin, 1944).

15. W.J.H. Sprott, *Human Groups*, (Penguin, 1967) p.31.

16. See for a report of current work Kline, Paul, *Personality Measurement and Theory* (Hutchinson, 1983).

17. The research is reported in Argyle, M. and McHenry, R., Do Spectacles Really Affect Judgements of Intelligence? *British Journal of Social and Clinical Psychology*, Vol.10 (1971), p.27–9.

18. Michael Argyle, *The Psychology of Interpersonal Behaviour* (4th edn., Penguin, 1983), p.106–7.

19. Katz, D., and Braly, K.W., Racial Prejudice and Racial Stereotypes, *Journal of Abnormal Social Psychology*, Vol.30 (1933), pp.175–93.

20. See for instance Scherer, Klaus R., and Howard Giles, (eds.), *Social Markers in Speech* (Cambridge University Press, 1979).

21. Kenneth T. Strongman and Janet Woosley, Stereotyped reactions to regional Accents, *British Journal of Social and Clinical Psychology*, Vol.6 (1977), pp.164–167.

22. Goffman, Erving, *The Presentation of Self in Everyday Life* (Penguin, 1971), p.60.

23. Brackman, J., The Put On, *New Yorker*, 24th June 1967, pp.34–73.

24. V.E. Mastin, Teacher Education, *Journal of Educational Research*, Vol.56 (1963), pp.385–6.

Further reading

More information about recent research on the topics discussed in this chapter can be found in:

Adler, Ronald B., and Towne, Neil, *Looking Out/Looking In: Interpersonal Communication* (Holt, Rinehart and Winston, 1984).

Robinson, Mike, *Groups* (Wiley, 1984).

Pervin, Lawrence A., *Personality. Theory & Research* (4th edn., Wiley, 1984).

SUMMARY SHEET

Chapter two – Audiences

Audiences, like crowds, have stronger emotions.

Write down the reasons for their presence.

What type of meeting is it?

How many people will there be in the audience?

Groups are made up of different individuals.

Find the group leader, and unite the group.

Audiences day-dream.

Use variety to help attention.

Attention (and day-dreaming) is involuntary.

People are less attentive during the afternoon.

Arouse interest by knowing their interests.

Enthusiasm helps to counterbalance the sleepy effect of sitting still.

Speakers are exposed: instead of ignoring this fact, try to understand how you relate to them, in age, status, job and personality.

You will be stereotyped: try to counteract the prejudice.

Performing needs careful stage-managing.

Watch the audience for feedback.

Don't run yourself down.

Feel sympathy for their attitudes.

3

Selecting, planning and arranging the material

Be prepared

If I were asked to give a one word explanation of the sort of confident, organized presentation we all envy, it would be *preparation*. The confidence comes from the speaker's knowledge that he or she has everything ready, has thought through the whole subject, and has enough of the right material to support the presentation. The sense of organization comes from the careful arrangements and selection of what is said, so that all the points are part of a logical order. Neither of these virtues are available to the speaker who bets on his luck (or cheek) and just talks off the cuff. Good speakers are prepared.

How do you achieve this? The whole of this chapter is about preparation. It is as much to do with the audience's abilities as the speaker's, and it is about the logic of organization, as much as the psychology of presentation. But the aim of all the advice is the same — that secure and admirable sense of being well prepared. There are two simple pieces of advice which start this process of preparation in the right way. Firstly, ask yourself what the *aim* of the talk is, rather than what the *subject* of the talk is. The first is much more specific than the second. If you plan to talk about a particular subject, you may feel the need to mention everything there is to know about that subject. But if the aim of the talk is to arouse the audience's enthusiasm for a research project on that topic, a brief sketch of the more exciting possibilities would be more relevant. A complete catalogue of every aspect will merely bore them, and will achieve exactly the opposite result.

There are many cases where the aim may be rather different from the subject. The advantage of thinking about the aim is also that then the decisions include the audience, and the audience's perceptions and needs, not just the speaker's ideas and knowledge. In practice, a very common mistake is to prepare a presentation as a speech on, for example, 'Heavy water reactors', without thinking whether the audience is interested in technical details or scare stories. If the *aim* is to reassure a local population that the heavy water reactor being built next to them is perfectly safe, then a lot of technical details about the design will probably scare them witless! Think of all your decisions when preparing the talk in terms of what you want the talk to achieve, and *not* in terms of what the bare topic of the talk is.

The second piece of simple advice is to prepare *more* material than you need. The idea of preparing 'just the right amount' is foolish. Until you start talking, you won't really know how much material you are going to get through, And if you insist on battling on to the bitter end of what you have prepared, you will almost certainly get the timing wrong, as well as turning the talk into a marathon. Talking should never be a dutiful forced march, it should always be an exploration, a discussion, a fascinating glimpse of the subject. It is an opportunity to learn about something new, which has to stop when the allotted time runs out. The best talks all end too soon, and the sense of having more to say, but having no more time, is the most satisfactory impression to leave.

The talk is also more interesting if the audience feel you are stepping smartly through the topic, summarizing far deeper knowledge and just mentioning the more interesting aspects. This impression is created if the speaker has more material than he needs at his finger tips; the need to summarize and curtail while he or she talks keeps up the level of tension, interest, and expectation. An audience should never come out of a talk feeling that the subject, like them, is exhausted. They should always be fired, rather than quenched. This happens best, if you prepare *more* material than you need. The habit of having extra material also allows you flexibility in timing when giving the talk, and helps you to answer questions at the end.

Thoughtful selection

The first important decision in making a selection of material is what you are going to leave out, not what you are going to put in. Talks are not for the transfer of a mass of information from one mind to others. That job is better done by paper. They are best used to give an overview of the subject, to create interest and enthusiasm. If you are already expert in a subject, you must now decide what the audience don't need to know; if you have to work on a new subject, as soon as you have understood it, you will have to make decisions about what is not needed in the talk.

It is a mistake to try to pad the talk out with masses of information and detail, in the belief that the audience will be impressed by your knowledge. They won't be – they will simply go to sleep. The amount of information which can be absorbed in one session when listening is strictly limited. The listening situation is quite different from sitting down with a book at a desk, and making notes. When listening, it is not possible to do more than gain an overall impression, and perhaps a handful of facts. Hard, dense packed information, cannot be communicated in talks; it is a mistake to try.

This fact is common experience – think how many actual details you remember from the last technical talk you heard – but it is encouraging to find it supported by research. Erskine and O'Morchoe did an experiment, in which they taught one class only essential principles with little detail, and then compared their knowledge with another class which had been given a lot of details. The first class did better. Their conclusion was that too much material causes interference, and the audience remember less, not more.[1]

Too much detail, then, is counter-productive in a talk (to use one of the familiar buzz-words of the 1970s). Factual information can only be used as illustration, or example, never as the substance of the talk. A verbal presentation communicates attitudes, enthusiasms, impressions, not facts. To try to battle against this natural situation will only alienate the audience, and reduce, not increase, the amount of information that is remembered. If you use cleverly designed visual aids, you may be able to incorporate a few figures and hard facts. But you certainly cannot expect the talk to be the source of reference for this information.

If you need to transfer a mass of solid figures, it is best to give a handout, with the figures tabulated for reference. You can then refer to a sample selection of the figures during the presentation, to illustrate the general point. But the aim of the talk should not be to learn detail. If you are talking on a technical subject, the audience should leave the talk with a desire to go further into the subject, or an impression of the range of complexity the subject embraces. They should not, and cannot, expect to walk out of the room with a mass of figures, facts, and details securely pinned inside their heads. Talks don't do this. Detailed learning has to be done with paper at a desk; talks are for interest and general information, not the transfer of a dense mass of information.[2]

The first task, then, is to select the material, and reduce the bulk of detail to manageable proportions. Selection, however, requires an aim, and this aim must be specific, not vague. It is impossible to make decisions about whether to reject, or leave in, a particular fact unless there is a very definite image of the audience and its aims in mind. So you must always select your material not for a general talk on the subject, but for a specific speaking task: for *this* audience, *this* task, and *this* amount of time. One consequence of this rule is that each talk you give must be considered separately. A general purpose talk will probably result in a vague presentation which will satisfy none of its audiences.

Another factor which must be considered when selecting information is the unloading rate, and the digestibility of what you are saying. People who are experts in a subject often fail to remember that it has taken them many years to get their minds around it all, and that what seems second nature to them now, may be confusing and alarming to a newcomer. The rate at which new information is offered is an important factor in the ability of the mind to absorb it. This factor, often not even considered, is so important that it is worth spending a little time on it.

Typically, experts assume the audience can absorb information *faster* than they actually can. I have rarely seen an expert making his subject too simple. So it is fairly safe to assume that you must introduce new ideas *more* slowly than you think necessary, and never more quickly. There are many techniques available to modify the rate at which new information is provided, You can, for example, modify the rate by repetition, example and anecdote. Simply repeating the same information in different words effectively halves the unloading rate. You can also open up more breathing space between ideas by adding new examples, which illustrate the same point, and you can provide a rest, while focusing on the same point, by including some amusing anecdote which relates to it.

One technique to ease the shock of new information is like getting into cold water by taking a wild plunge. It works by giving the audience a full list of the topics and key words at the beginning of the talk. This is the sudden plunge, and they will then need reassuring that it is not as frightening as it sounds. You can then go back to the beginning, and start again with the first point, slowly making it clear.

It is worth spending rather longer on the first point, giving lots of examples and supporting information, because if the audience can be made to understand the first point, they will approach the rest with more confidence. By shocking the audience with their inability to comprehend the whole subject, and then proving to them that they can, after all, be brought to understand the first point they come to, you will boost their confidence in their ability to learn.

Mix old and new

Another way of reducing the unloading rate is to ensure that there is a mixture of old and new information. Some speakers seem to think that they must retail only new facts, and can ignore the old facts. This is not so. The old facts are the foundations on which the new facts must be built. These foundations will be buried under all the other daily information the audience must cope with. You must uncover, bring to light, or remind the audience of what they already know before adding new information. It also controls the overall unloading rate. A mixture of familiar facts amongst the new reduces the total strain on memory and comprehension. It gives the audience a satisfying feeling of competence. The feeling of smugness, in the unspoken reaction 'we *know* that', will transfer to a feeling of interest and respect if it is followed by the reaction, 'but we didn't know that'. If, just when the feeling is becoming, 'we can't cope with all this', you introduce more familiar material, the audience will feel themselves on firm ground again. By alternating familiar and strange, new and old, the audience's comprehension is kept flexible and alert.

The technique of mixing familiar and new is supported by theorists of communication. Umberto Eco makes a technical point about the communication of information, which confirms this important principle in selecting information. The content of a presentation cannot be all new; some of it must be familiar, even repetitious, in order to orientate, and rest, the listener's mind. Eco insists that there must be:

> communication dialectic between probability and improbability (that is between the obvious and the new – and ultimately, in a more technical phraseology, between meaning and information). A high rate of improbability runs the risk of not being received, and therefore the message must be tempered in small degree with conventionalities, commonplaces, and must be reiterated . . . One

of the problems in message-coding is the balance between the obvious and the new. How few conventionalities are necessary to communicate a piece of information (as a new thing?).[3]

To achieve this controlled unloading rate, with a mixture of familiar and new information, you must carefully select the examples and analogies. Of course, it is not possible to give a formula for the exact unloading rate appropriate for a particular audience, or to provide an infallible rule so your presentation will be just right. But this doesn't matter. What matters is that you have thought about the problem, and are aware that you must watch the rate at which you put out new ideas.

Judging the selection of material is more a matter of conscious awareness, than of perfect correctness. Thinking is what matters; don't blunder on oblivious. Audiences are flexible, subjects have many different ways in which they can be presented, there is usually a willingness to learn in the audience, and an ability to figure it out for themselves. The only rules are to remain aware of audience reaction as you talk, and be prepared to modify what you are saying if blank incomprehension, or glazed stares of boredom, meet what you have said so far.

Vivid and entertaining examples are often the best way to engage an audience's attention, and to ease the passage of new information. But this does not mean that you should load example after example onto an already satiated audience. Avoid indiscriminate use of all the examples you can think of; choose only the best ones. The examples and analogies you do use must be brief, familiar and concrete. It is often difficult to think of good examples, and one writer on the subject, Donald Bligh, admits that, 'Personally I find that I can never think of good examples at the time of lecturing. They therefore have to be prepared in advance. In fact it is quite a good idea to collect examples at all times.'[4] Following the second principle of this chapter, it is a good plan to have *more* examples than you need, and to make a selection as you talk, depending on which type of examples seems to strike a sympathetic chord, and how much relaxation, or increase, of the unloading rate the situation requires.

There is no doubt that the best way to make the talk memorable is to use materials which are specially relevant to the audience, dramatic, or simply funny. But this can go too far. One speaker, a most distinguished man in his own field, tried to make his lectures memorable in an unusual way. Gray Walter, the famous neurologist, 'asserted that essential ingredients of a successful lecture were humour, horror and sex. To provide for this alleged desire for sensation, in an erudite lecture on 'Brain mechanisms and learning' he used coloured backgrounds for tabulated data in which such outlines as bathing beauties were engraved in white.'[5] The problem with such tactics is that the audience can sense when they are being pandered to, and soon resent it.

To set out to entertain before anything else will not only fail to communicate the necessary information, it will probably lose the respect of the audience as well. The ideal, as in everything to do with speaking, is to provide as much change and variety as possible. So mix theory with anecdotes; and mix humour with serious points. If you have just given a dry, detailed and strenuous exposition, lighten the atmosphere with an amusing story. And if you have just told a long anecdote, take the opportunity to emphasize a complex theoretical point immediately afterwards. In this way the audience is kept alert by the ever changing demands being made on their attention.[6]

A coherent pattern

An early decision you must make when preparing a talk, is how are you going to organize the material. Read through the notes you have gathered, selecting what is most useful, and considering the best arrangement. No talk which is nothing but detail from beginning to end will have much permanent effect; nor will a talk which appears to make only one point. Research shows quite clearly that the listeners remember better, and remember more, if they have a sense of the *shape* of the talk. Any subject can be broken up into separate points. Even if your talk is about one chemical reaction, say, you can break it down into an overview, the raw materials, the theory of the process, the construction of the reactor vessels, the control and supervision of the reaction, the discharge and customer delivery problems, and a general summary of the process. In this way, one subject becomes several points. The listeners must be able to grasp the structure of the talk: make sure that you make the overall *pattern* of your presentation plain to the audience.

The only way to make a pattern plain is to make it bold. Since a grasp of the pattern is so important to a satisfactory sense of understanding, the best technique is to make each individual section, as well as the overall pattern, simple, logical and clear. Most speakers fail to realize just how strong and stark this pattern must be. They forget how much contrast is needed to make the picture stand out. Because they are familiar with it themselves, they do not realize that the audience, not quite awake anyway, may find a new subject confusing.

I heard one talk which had been a confusing drone of detail: but when asked the speaker insisted that his notes were clearly structured into five different sections. He had simply failed to make it clear to the listeners when he was moving from section to section. It was clear enough to him – he had his notes in front of him. But the audience couldn't see the notes, and were mystified every time the subject seemed to have changed without warning. It is almost impossible to make the structure too clear: the listeners need to grasp the shape, pattern, or structure of the talk so that they have a framework to hang all the details on. Unless they perceive this structure, they will be left with a mass of shapeless details. The job of the speaker is to make the framework clear. It is difficult to overdo that job. Fig. 3.1 shows the sort of framework the talk can be built on.

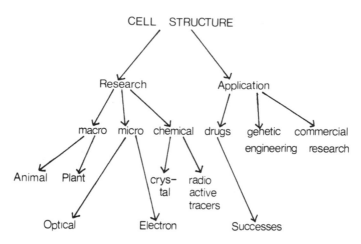

Fig. 3.1 Framework for a talk

The choice of pattern is the next important decision to be made in preparing the talk. Ask yourself whether the arrangement of information you have chosen has a discernible pattern. If so, is that pattern suitable for the type of audience, the type of talk, and the subject matter (in that order)? Most important, will the audience perceive that pattern? It is very difficult to understand, when we have seen a pattern ourselves, that others may not be able to see it. Yet there is evidence that seeing a pattern is something which is learned, not something inevitable. For example, when the Pygmies in the Congo jungles were first shown black and white photographs they could see nothing but an abstract pattern of black and white blobs. They had to learn to perceive this flat, apparently random sheet of blobs and smears as an image which imitated the real three dimensional world. Your audience may be like the Pygmies in their understanding of the pattern of organization in your presentation. Make sure that you explain the structure of the talk, so that lack of familiarity will not make the listeners see your talk as a random maze.

Psychologists have done much research on how we perceive patterns, and what sort of patterns are most easily understood by human minds. It is worth pausing for a few pages, to look at this work, because it illustrates very clearly what the speaker must do if he or she wants his audience to understand. I have chosen four topics in the perception of patterns to start you thinking about the kind of organization a talk needs. The first of these is Gestalt psychology, the second is the importance of our sense of place, the third is the significance of patterns of seven elements, and the fourth is the way we chunk details to make them easier to understand and remember.

The first topic I want to explore is the theories of the Gestalt psychologists. A group of early experimental psychologists working in Germany, the Gestalt psychologists, showed in a series of elegant experiments that human beings naturally see patterns in the objects around them. This tendency was universal, and was a powerful element in human understanding of reality. We grasped the mass of detail in the real world, only by seeing it as part of a pattern. The Gestalt psychologists formulated laws for the type of patterns which were most easily perceived by the human mind. It is wise to follow these rules in the construction of a pattern of organization, because they define what will be seen as a pattern, and what will be seen as a jumble. Max Wertheimer suggested five laws – similarity, proximity, closure, good continuation and membership character – which the mind used to impose order through pattern.[7]

Obviously, patterns can be most easily made from things which are similar, and close together. Five trees of the same kind, standing in a group on the sky line, will always be seen as a pattern. So will three blue cars of the same make next to each other in a car park. And so will the three advantages of several different kinds of software, when balanced against the three disadvantages of each kind. Closure, and good continuation are more technical ideas. Human beings want things to be complete, and we will often make the completion ourselves if it doesn't exist in reality.

A drawing of a circle which has a gap in it will be seen, and remembered, as a complete circle. Similarly, if the talk about software only mentioned *two* disadvantages in the third of the types, most people in the audience will remember three disadvantages for each type. They will even invent the missing disadvantage, or transfer it from another type, to make a complete pattern of disadvantages. Good continuation is a similar requirement in patterns. If a line is continuous, we can see it as a pattern more easily than if it has large, random gaps. Usually, we fill in these gaps ourselves. Thus the line in the middle of the road is seen as a line, even though it is half bare asphalt surface, and half short white dashes.

Finally, membership character means that it is easier to perceive a pattern if the objects all seem to belong to the same group. Thus it will be easier to remember a group of points if they all belong to the same topic. A talk about five important engineering principles in bridge design, with one section on the power output of modern engines, would fail on this principle. Most of the audience would perceive the section on engines as a digression; they would remember only the five principles, and probably also remember that there were some red herrings in the talk. If, however, the talk had been presented as six principles of engineering, applied to metal structures, such as bridges, engines, and steel framing, it would have been grasped as a pattern.

The results of Gestalt psychology are widely accepted. The ideas can be applied to ensure that the organization of a talk has a pattern which is easy to grasp. If the structure of the talk fits in comfortably with the natural way the human mind grasps patterns, it will be better understood and better remembered. Try to use the principles in laying out your talk using similar sections, clearly part of the same topic, and all related to and continuous with each other.

Mental orientation

The second topic is the importance of place, and our sense of location. Doubtless because human beings evolved from organisms for whom finding their way around was vital to survival, we have a highly developed sense of location. The physical shape of the environment is usually the first thing we grasp. Many people buy a map soon after arriving in a new town, so they can satisfy this need to locate themselves. This sense of location is strong, and patterns which can be visualized as a plan are the easiest to grasp. Topographical arrangements were used in classical times as props for the memory:

> In the Roman system each item to be remembered was associated with a particular place in the surroundings. A person would first form a standard list of locations that he knew well. Maybe he would take his courtyard and make the pillar on the left the first locus, the ledge next to it the second locus, the tree the third locus, etc . . . Each of the topics to be remembered was associated with one of the standard places in the person's system . . . The remnants of this system have descended into our own speech as 'In the first place . . .' and 'In the second place . . .' The Greeks used a similar system of linking ideas with places. Our word 'topic' comes from the Greek 'topos', which likewise meant 'place'.[8]

It is a story every speaker should bear in mind. If the audience become disorientated – the physical metaphor of location is used again to describe mental experiences – they will neither understand nor remember. If you can give the talk a physical shape – for example by going through a process from the factory gates, then the processing plants, to the loading bay – it will be easier to grasp. One of the reasons why writing for the computer industry is often poor is that spatial metaphors are mixed up, and no clear sense of location is given to the sequence of ideas. Walt Disney, in *Tron*, used the primitive sense of adventure, journeying, and locations, to make abstract ideas comprehensible. While every talk doesn't need to be like *Tron*, using simple shapes, like a curve of points, or a tree of branching possibilities, will help the talk to be comfortable, not confusing. I have said already, but it is worth repeating, that the structure must be an order of simplicity clearer than you think. You know where you are going; the audience are new to the area. They need very clear maps and signposts to avoid getting intellectually lost.

The sense of being well guided will always have a beneficial effect. Even if the pattern is not strictly logical, the appearance of logic will soothe the audience, and give them confidence in their ability to understand the presentation. Bettinghaus reports research to show that:

> There seems to be some merit in having the appearance of logic . . . What seemed to be operating was a feeling by audience members that 'logicality' ought to be a desirable thing even though there was

no evidence that any of the audience members could tell logical appeals from illogical appeals.[9]

If the talk has a shape, so that listeners know where they are all the time, the actual logical connections can be tenuous. It is the *sense* of knowing where the speaker is going that matters: nothing is more unsettling to an audience than a mystery tour.

The magic number

The third of the topics in psychological research on the perception of pattern, is the importance of the number seven. In a memorable paper the American psychologist, George Miller, describes research on the brain's ability to perceive, distinguish and remember. The argument is detailed, though highly readable, and I recommend Miller's essay. After discussing a variety of experiments, Miller writes:

> If I take the best estimates I can get of the channel capacities for all the stimulus variables I have mentioned, the mean is 2.6 bits.

'2.6 bits' refers to the binary representation of the decimal number seven. He suggests that:

> We are therefore in a position analagous to carrying a purse which will hold no more then seven coins – whether pennies or dollars. Obviously we will carry more wealth if we fill the purse with silver dollars rather than pennies. Similarly we can use our memory span most efficiently by stocking it with informationally rich symbols.

Miller ends his essay with a peroration on the title, 'The Magic Number Seven':

> What about the magic number seven? What about the seven wonders of the world, the seven seas, the seven deadly sins, the seven daughters of Atlas in the Pleiades, the seven ages of man, the seven levels of hell, the seven primary colours, the seven notes of the musical scale, and the seven days of the week? What about the seven point rating scale, the seven categories for absolute judgement, the seven objects in the span of attention, and the seven digits in the span of immediate memory?[10]

The message is clear; try to find about seven main sections in your subject ('plus or minus one or two', as Miller says) in order to make it convenient for the audience to remember the general shape of the presentation. It is usually easy to find seven sections in a talk – the overview at the beginning, and the summing up at the end count as two sections. Most topics can be broken down into another four or five sections. The important point is that too many sections are worse than no sections at all – the audience will feel they ought to remember them, and will fail. The evidence is that somewhere between five and nine major headings is right. If you have eleven headings, group them into smaller numbers, if you have only three, try to split them. With a little imagination, any subject can be divided up in various ways, and it is up to the speaker how to group the points to make a clear, and memorable, pattern.

It would be futile to give templates for organizing a talk, beyond the advice to have about seven sections. There are various principles of organization, and the disadvantages and advantages of each one depend so much on the subject, and the aim of the talk. The evidence for the case you are making can be arranged hierarchically, by classifications and groupings, by a chaining technique, by comparisons, or by networks. Another method of organizing a talk is to base it on a developing idea. The development can be from the general to the detailed; from the key detail or arresting example on to a general point; from a brief summary of the conclusion, followed by detailed arguments; or you can work up to the point, only revealing it at the end of the talk. Any of these structures will give a pattern to the talk. All that matters is that the audience can see, and understand, the organization of what you are saying.

As with many other aspects of spoken presentations, research has been done on the merits of different patterns of organization, but the results are inconclusive. It is best, therefore, to choose a pattern that seems to fit naturally into the material, or that springs readily to mind. What matters is not which pattern you choose, but that it should be explicit.

Chunking

The fourth of the topics, which illustrate current knowledge about the way the mind perceives patterns, is the importance of 'chunking'. Chunking is the way the mind collects complex objects into groups, so that it can grasp a large amount of detail, via a few more general divisions. Chunking is a vital element in memory. Peter Russell explains:

Chunking is so natural that most people do it anyway without realizing it. Given the number 572317482, a person would probably regroup it as 572,317,482 . . . In doing this not only has a nine-figure number become three groups of three, but the individual digits have been given different tags – 572 has become 'million'.[11]

Another way in which facts can be systematized is by remembering a principle, rather than the details. The speaker can use this technique to help the audience increase their grasp of the subject. If the principle is understood, the details can then be reconstructed from the principle. Russell gives a good example:

Following are three lists of ten numbers (each therefore longer than the average immediate memory span of seven items).

(a) 0 1 2 3 4 5 6 7 8 9
(b) 8 6 4 2 0 9 7 5 3 1
(c) 1 8 4 5 7 2 0 9 3 6

The first list is easy; the sequence is already well known . . . The second is still moderately easy, once you see the principle under-lying the sequence. . . . The third is more difficult because the pattern is not so obvious. . . . The rule is: Add 3 to the first to get the third, and 3 to that to get the fifth, and so on for every number. Subtract 3 from the second to get the fourth, subtract 3 from that to get the sixth, and so on. Ignore any '1111' digits . . . Got it?[12]

Both chunking, and a clear principle of organization, then, help the mind to cope with details. If you organize your talk using these principles, it will be easier for the audience to understand, and therefore both more enjoyable, and more memorable for them. There are thus four different kinds of psychological evidence, which give the speaker insight into the best way to structure the talk, and the ways the audience will perceive a pattern in what he or she says. Firstly, the work of the Gestalt psychologists underlined the powerful pattern making and pattern seeking abilities of the mind. Secondly, the spatial orientation abilities of the mind can be used to give a shape to the talk, by structuring it around locations. Thirdly, there is evidence that around seven is the right number of sections to use, and fourthly, chunking details helps the all too human listener to grasp and remember what you are trying to tell him.

Before leaving the subject of clear organization, let me make two more general points about the way you structure your material. The first is that the whole carefully built edifice will come tumbling down if you make mistakes. The point of organization is to make the subject clear, and to build the audience's confidence that they are on a worthwhile journey, not a desultory ramble. To preserve this clarity and confidence, you must always be careful about the definition of terms, the evidence, the alternatives, the sources, and the reliability of your material. You do not necessarily need to specify all these details while you are talking, indeed to try to do so may make your presentation hopelessly complex. But you must have this information available in case you are asked. One disputed fact, one palpable exaggeration, one proven mistake brought to light by an unfriendly questioner, can wreck a whole edifice of carefully built argument.

The second general point is that a talk is a dynamic, progressing experience, not a static structure. It helps the audience if they see the structure of the talk growing before them. So instead of just showing a list of the sections, build up the structure dynamically. The way to do this is to have a section of the board, or a flip chart or overhead projector slide, on which you write the new heading each time you move to a new section. Alternatively, you can use a cover sheet gradually moved down an over-head projector transparency to reveal new topics one by one. Research shows that listeners are much more likely to note and remember key words which are written on the board, flip-chart, or overhead projector. This may partly be because they have time to absorb the information, and have a micro-break while the speaker pauses to write.[13] But it is mainly because they are alerted, each time, to the change of topic, and can look back and see how it fits in with what went before. The audience gets a sense of a structure growing dynamically, and the extra confidence adds greatly to the success of the talk.

The structure of reasoning

The third job of preparation, after selecting the relevant material, and organizing it in a clear and simple structure, is to form the details into a coherent argument. Every talk has a case to argue, unless it has no more structure than a telephone directory. The way this case is argued, the way the details are marshalled as evidence for the points being made, is an important part of the planning.

The first point is that you cannot make the argument complete, and prove every point and detail. Spoken argument can only sketch the outlines of the case. It gives an emphasis and immediacy, through personal involvement, and immediate feedback from the audience, but it does not give a forum for the fine points and the mechanics of scientific proof. Speaking can only offer the bold outlines of proof, not the inner workings, of the argument. The details are better written down. So if you have to present a case which has a detailed proof, you must summarize, give a few salient details, and refer your audience to the published papers, or internal reports.

Skimming over the details of reasoning does not relieve the speaker of the need to be accurate. It is wise to avoid apparently spurious argument; the audience will spot it. Ogden and Richards warn about the 'process of 'lubrication', the art of greasing the descent from the premises to the conclusion.' While such a process may seem attractive, you should not imagine that the audience is less sharp, just because they have fewer details. They have more time to think, and will be able to bridge the gap between generalizations, to work out the details for themselves. Unless the story hangs together, they will not believe what you say. Accuracy is just as important in spoken presentations, as it is in published papers.

The art of conveying an argument in a spoken presentation lies, then, in the selection of the examples to use. Most scientific arguments are based on induction. But you should never try to offer complete induction, listing tables of figures and results. In a talk you can only state the hypothesis, use a few illustrative examples of the kind of results obtained, and state the conclusions. Of course, you must make sure that the examples you give are typical. One way to do this is also to give exceptions to validate the examples, and forestall criticisms.

The old adage is: 'The exception *proves* the rule'. This saying, incidentally, is usually misunderstood, because 'proof' means *test*, as in 'proof' spirit, or 'proving' a gun. The adage means that the rule is checked by the exception. Can the exception dent or explode the rule you are trying to prove? Will the rule survive? The purpose of giving exceptions is to see whether the exception can be explained away as an irrelevance, not a true exception, or an example of some other rule altogether. If not, then the rule must of course be modified to take account of the facts which it does not satisfactorily explain.

The psychology of audiences requires that an attempt is made to illustrate a rule, or general conclusion, with obvious and typical facts, as well as apparent exceptions. The speaker's job is to allay doubts, and calm suspicions, not to produce cast iron proofs. He has also to explain and clarify the rule, and exemplification is often the best way of explaining, as well as the best way of justifying, a conclusion. But as speaker you should never lose sight of the role of examples as aids to the audience, rather than as elements in a scientific method. If you keep this distinction in mind, you will be able to persuade yourself when preparing the talk that not all details, facts and figures are relevant to the task. A carefully selected set of illustrative examples is all that the structure of the reasoning requires in verbal presentation.

In considering the type of examples to choose, the speaker must find what will be easiest for the audience to understand in the limited time available. Thus, there is evidence that human beings prefer direct proof to indirect proof. That is to say they prefer to be shown that something *is* the cause, not that all the other possibilites can *not* be the cause. This seems to be because of the universal tendency to cognitive economy: in other words people like the simple and direct route to a conclusion, rather than one which requires sustained attention, a sharp memory, and active deduction. So the wise speaker will construct his argument as a direct, not an indirect, proof. The aim should be to give strong, simple arguments, which offer clear, uncomplicated reasons, not elaborate and intricate analyses.

Chaining

Not all talks fit into this pattern of a clear point, a few strong examples, and a conclusion. Sometimes there is no argument, simply a story to be told, or a sequence of generally interesting facts to be repeated. A popular form of organization for this kind of talk is called chaining. Figure 3.2 illustrates it. It is used most often in children's stories, but also by journalists, who rely on continuing curiosity, rather than memory or argument, to maintain interest and attention. Alistair Cook is a good example of a journalist who pieces his items together by chaining. The technique can often be seen in lead stories, as well as general interest broadcast talks. It is a useful way of giving shape to a series of loosely related points. The problem with the technique is that if the audience's attention wanders, and a link has been lost during the period of inattention, it will be difficult for them to see how the present point in the chain relates to previous points. It is also more difficult to remember the whole chain, since the structure of the talk only provides links between any two adjacent items. For this reason regular summaries are needed.

There is one more important warning to heed in any reasoning structure that you design for your presentation. It is most important that the problem is clearly grasped before solutions are proposed. The rigorous laws of induction and deduction apply to the laboratory work, but not to the spoken presentation of that work. The art is to try to give a goal-orientated speech. First define the problem, then survey the solutions, then choose an answer which is a valid solution

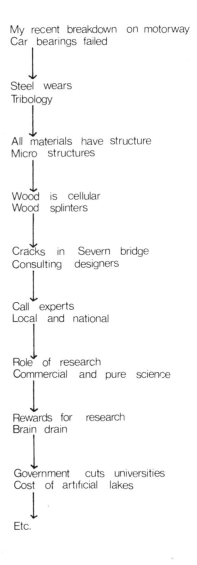

Fig. 3.2 A chaining organization

to the problem. Unless this is done discussion of the proposed solutions, and the arguments over their merits, will be meaningless. The audience will discard them as so much background noise, waiting for something which has a recognizable purpose. Make the problem a focus, and keep on referring back to it, showing how each point or section relates to it. You should be taking stock of where the effort towards a solution has got to all the time. Experiments show that an audience does not make 'even the most important and simple inferences' when listening. It must all be made clear by the explicit structure of the talk.[14] The speaker has a duty to make the development of the ideas coherent, and transparent to the audience. Unless the structure of reasoning is lucid, the audience will feel it is blindly wandering through a maze of information.

To conclude this section on the presentation of reasoning, let me repeat what I said earlier. The evidence is that which structure is chosen does not matter, so long as there is some perceivable structure. When Beoghley (1954) changed the order of certain paragraphs in a lecture he found no significant difference in the student's retention of the information. Does this mean that organization does not matter? Clearly not. What it means is that there must be a structure, and it must be clear. But which of several different possible structures is chosen is less important. It is better to spend preparation time deciding how to emphasize the structure, than in agonizing over which particular structure to use. Thus, you can plan the transitions from point to point in the structure of the talk, using such techniques as changing your position, using a different tone of voice, pausing between topics, or simply cleaning the board to mark the change of topic. Time spent on these details will show greater benefits than time spent choosing a marginally better organizing principle. Clarity and simplicity of structure will triumph – the fine print of scientific evidence is for writing, not for speaking.

In search of being well remembered

The fourth job of preparation, after selection, organization, and coherent argument, is to consider how you are going to make the presentation memorable. By this I don't mean an earth shattering and unforgetable experience, but simply how are you going to make sure that the audience carry away something useful? Clearly, an informative talk should not be ten minutes fun, with nothing to show for it afterwards. The pleasure is important enough: all useful things are accompanied by pleasure, and mental pleasure is an inescapable sign that learning is taking place. You always remember what you enjoyed. Pleasure is a vital biological indicator of what we should be doing; if there is no pleasure, we shouldn't be doing it. Of course, you want to make the talk enjoyable. But you want to do more than that. You want the audience to have lasting benefit from it.

How do you ensure that the talk is easy to remember? This section will explore, briefly, our understanding of memory, and how memory works. Research this century has uncovered results which can be applied to preparing a talk which the audience will find easy to remember. The first thing to realize when planning to be remembered is that the speaker has a very different problem from the writer. I can't show in a book the different effect of listening to words, rather than reading them. But heard language is remembered in a very different way from read language. Read language uses the very powerful visual memory; heard language uses only the auditory memory, which is less exact. Notice how often you are not quite sure what you heard. How often do you remember exactly what was said, rather than the gist of the conversation? But we can usually remember vividly what we saw.

Research on memory has shown that there are two different kinds of memory; the short-term, and the long-term memories. The short-term memory typically stores the individual words and phrases used. The long-term memory is semantic, and remembers ideas and facts, not words and phrases. There is evidence that what we remember from a talk is the kernel of ideas, plus a vague sketch of the words it was encoded in. If asked to reproduce the information people will recode the message, supplementing their memory of the facts with guesses about the words they were encoded in.

The long-term memory is permanent, and remains despite interference from other sources. Psychologists define short-term memory as one minute, although anything up to 20 minutes may be erased by powerful events. There is no sharp divide between short and long term memory; one is gradually converted into another over a period of up to half an hour. A bang on the head, such as a fall or a car accident, will often erase memory of the few minutes just before the event, although memories which have become long-term will be retained.

The speaker can make use of this information to design his talk for maximum memorability. He cannot expect individual words to be remembered for more than a minute or two. He or she must also expect what he said to be erased by the events of the next hours after the talk. The speaker's best chance is to choose a few key points, and reinforce them so they become part of the long-term memory. Up to six or seven chunks of information, grouped under keywords is the best tactic.

New theories

Recent research has offered two new theories about how memory works. They give us new insight into our own memories, and into what we can expect from an audience's memory. These theories are the 'protein' theory, and the 'holographic' theory. The protein theory suggests that memory works by encoding short-term electrical changes into long-term chemical changes in the brain. Researchers found that it was possible to transfer memories from one animal to another by injecting extracts from rat brain into another live rat's brain. Researchers were able to isolate individual protein chains which seemed to be responsible for individual memories. There are many billions of different proteins, and if such a mechanism is working in the human brain, it might explain why long-term memories form slowly. It might also explain why long term memories are of simple, global facts, not details and figures. There is also evidence that these proteins last for only a few days, and must be resynthesized to keep the memory traces strong. Regular revision, and repetition is needed to preserve strong memories. So if the speaker is giving a series of talks, or a weekly seminar on a topic, he will help his audience if he spends a few minutes each time summarizing the previous talks, and reminding them of the main points of the series.

A second theory of memory hypothesises that memory patterns are distributed throughout the brain in the same way that the wave forms of light are distributed throughout a holograph. This theory helps to explain why very large numbers of different memories can be stored without interfering with each other. The holographic theory may also suggest why the same experience can recall similar memories which have been lost. The illumination of a holographic plate by the same wavelength of light as that which originally recorded the image reveals the original image. In the same way, Proust's Madeline cake evoked his whole childhood. Almost everyone has experienced a situation where the repetition of an event hooked-out whole chains of memory.

These theories illuminate the working of our memory. They also remind us of the way this material must be presented and reinforced if we want our audience to remember it. A talk which is merely a passing entertainment, but which leaves no permanent memories, is of little value. Yet many speakers are so worried about getting through the ordeal, that they never think about the long term effects of their talk. Their ambition stretches no further than getting to the end of the talk without disaster. Yet most of us would not be satisfied if we knew that everything we said, and prepared, would be wasted. Most people want to be remembered, and some understanding of how memory works helps you to plan for this end.

Memory research

The earliest research work on memory was concerned with the permanence and reliability of memory. The father figure of memory research, Hermann Ebbinghaus, researched the ways in which memory traces decay. He showed that there was a rapid decrease in the amount that was remembered in the first hours. This was followed by gradual tailing off over the following days and nights. But most forgetting happens straight after the learning. Even one hour afterward, more than half the original material was forgotten. Nine hours later about 60% had gone, and one month later 80% had been forgotten. Many experiments since have confirmed this 'curve of forgetting'. But recent research has also shown that the more organization, significance and associations are built into the material, the more gradual the forgetting. Research has also shown that for a short time after the learning session, the amount remembered may actually increase slightly, before beginning to decline along the general curve. This has been christened the 'reminiscence effect', and the better organized the material, the stronger is this effect. [15]

These general results of memory research are well established, and the speaker must take account of them. There are three major lessons to be applied to the preparation of talks. Firstly, that most of what you say will be forgotten, and therefore you should chose carefully the points you want to be remembered, and concentrate on those. Secondly, that clear organization, a recognizable structure, is a powerful prop for memory. It acts like a scaffolding to hang ideas on, and is the main way you can make a talk memorable.

The third lesson deserves rather more consideration. It is that the period between an event, and the memory, is significant. Memory increases briefly after an event – the reminiscence effect – providing it is not interfered with. Silence and pauses act as reinforcement. Fresh stimulae tend to partially erase the memory of immediately preceding events. The main technique to use is to repeat the important points at intervals. Memory must be reinforced; arrange for the information to be processed at least twice by the mind, preferably in different forms, such as auditory and visual. This reinforces the semantic memories. Psychologists know that muscular, visual and kinaesthetic (the sensations of movement) experience all help to reinforce memories by providing alternative ways for the memory to be encoded and stored.

The need for pauses to reinforce memory is only one of the ways in which the spacing of events affects the way they are remembered. There are three well-recognized effects on memory: the 'Von Restorff effect', the 'primacy effect', and the 'recency effect'.

1. *The Von Restorff effect.* This describes the way in which we remember unusual events. The point at which a speaker drops his glass of water will be remembered, although everything he said may be forgotten. It is the unusualness of the event which makes it memorable. You do not need to drop glasses of water to be remembered, though. A speaker will increase the chances of important items in his presentation being remembered if he makes them outstanding, startling, or in some way memorable. Simple acts like writing a key-word on the blackboard, flip-chart, or over-head projector slide, help to make them stand out. Associating a point with a funny anecdote, making an unusually emphatic pause before and after the point, or even simply changing position to a different side of the table, increases the chance that a point will be remembered because of the Von Restorff effect.

2. *The primacy effect.* There is an increased probability that people will remember the first two or three items in a talk. This effect also operates after any break or pause, such as a coffee break, or even a short pause to set up visual aids. Even slight variety of position, or presentation, will alert the audience, and what you say immediately afterwards will benefit from the primacy effect. As the talk progresses, new information overlays the first information, but the memorable effect of the first few things is never entirely lost.

3. *The recency effect.* This is broadly the inverse of the primacy effect, and arises because we tend to remember the last thing that happens. When listening to a talk, the last point which is made is likely to be remembered rather better. These effects are slight, and other factors, such as emphasis, organization, and variety will also influence what is remembered. But in general, what is said at the beginning of a verbal presentation, and what is said in the concluding minutes, are better remembered. Similarly, if there is a pause in a presentation, the facts

just before and just after the pause will be better remembered.

Breathing space for memory

During the body of the talk, it is too easy for ideas to become jumbled together in the listeners' minds. Unless the speaker is careful to mark the sections with clear signposts, the landscape of the talk will merge into a blur, and few details of the mental journey will be remembered. Where new ideas interfere with ones previously presented there exists what psychologists call 'retroactive' interference. Where old ideas interfere with those following (perhaps because they are so arresting that the listener's mind is still partly dwelling on them) psychologists call it 'proactive' interference. The technical terms are not important, but the principle is. Memory depends on clear space around important ideas and facts, and rest or silence is the best form of clear space.

Despite every care, it is still not possible for a speaker to ensure that everything he says is remembered. Partly this is just an inescapable fact of mental life. People forget. Forgetting itself is an interesting subject and psychologists have suggested a number of reasons for forgetting. They include repression, a word coined by Freud to describe the way people force out of consciousness events and ideas which for some reason they do not want to remember. Motivation is as important in forgetting as it is in remembering. Quite slight motives can lead to forgetting, such as an embarrassing professional lapse, which may lead to all information surrounding the situation being repressed.

Other causes of forgetting are the decay of the memory traces for physical reasons, interference from other similar memories, and loss of the ability to locate the memory. According to this last hypothesis, all memories are retained in the brain, but retrieving them is too difficult. Recalling other features of the experience, or working gradually back into our memories, will often enable us to discover memories we had thought lost. Thus my father, writing his autobiography for his grand children in his seventy-second year, started re-telling an incident in which a school-fellow had died of meningitis. To his surprise he found he could remember the boy's name, and even his address. He had not remembered those facts for sixty years, and certainly could not have recalled them had he been asked. The 'search' theory of memory claims that forgetting happens when more and more memories are built up without enough features to differentiate between them. Unless simple clues are given to act as handles, it becomes harder and harder to find any particular fact, from the mass of detail in the memory.

Let me make three final points about memory before concluding this section. The first point is that it is useful to distinguish between active and passive memories. Faces seen at a meeting, for example, remain familiar if we meet them again, even though we may not be able to place a name on them. The image of almost every face we see is retained in the memory; but voluntary access to the information has been lost. Recognition is passive; active recall requires a pathway into the memory. Effective memory relies on clear structures being developed to retain access, and not just on the vividness of the memory itself. The speaker's task, therefore, is to provide this unforgetable structure, without which the detail amassed in the talk will be lost, like water poured onto sand.

The need for organization to help active memory leads me to the second point. Unless we can see a structure in the details, they are less meaningful, and therefore much more difficult to retain in memory. Donald Bligh gives a good example of the effect of meaningfulness on memory.[16] He asks his reader to attempt to memorize the following lists of letters:

yma	try	the
uyo	can	lid
ont	but	may
teh	you	not
dil	yet	fit
ryt	not	yet
ubt	the	you
nac	may	can
tey	fit	but
tif	lid	try

It is immediately obvious that learning the first list would be a major task, whereas the middle list is only moderately hard, and the last list is perfectly easy to remember. The conclusion for the speaker is obvious; break up similar facts into patterns. Remember that you must link new ideas into existing ones, using a clear structure, if you want your audience to remember them

The third and final point about memory is a more hopeful one. It is a strange fact that we are usually modest about our memories. Hans

Eysenck comments that:

> when you ask most people about their memory, the first thing they usually say is that they have the bad luck to have a very poor memory. There is an interesting contrast here with what happens when you ask people about their intelligence or sense of humour: only a very small percentage of people will admit to below-average intelligence or a poor sense of humour![17]

The memory is extraordinarily powerful, much better than we think. When tested on their memory of ten thousand pictures, people recognized 99.6% of them correctly. As the researcher commented: "the recognition of pictures is essentially perfect".[18] The brain is highly sophisticated and memory itself has no visible limits. The audience could remember much more than they often do; the amount is not limited by any natural maximum capacity. If the speaker prepares his talk in a way which provides the opportunities for memory, and offers a clear scaffolding of organization on which the memories can be hung, there is no limit to what the audience can be helped to remember.[19]

Preparation is half the battle

Preparation, as we have seen, is an important part of the success of the talk. Speakers who think that they can throw together a few facts and figures, and rely on their instincts to present them interestingly, often have a sad shock. Their timing goes wrong, their information gets jumbled, the audience gets bored, and remembers nothing from the presentation. Worst of all, the breezy confidence and charm which was going to carry the day, ebbs away as the chaos becomes obvious. Decidedly, to try to talk unprepared is inviting disaster.

Preparation should consist of careful working through a list of points, checking that they have all been considered. Is there a comfortable surplus of material in case of need? Has the material been carefully weeded, so that only the most striking, interesting, and relevant examples are used? Is there a mixture of familiar and new material? Is the organization of the talk clear, and does it form a pattern which will be easy to grasp? Have you chosen about seven major headings, are the details collected into recognizable chunks, and have you planned to present the progress through the talk dynamically? You must also consider whether the organization follows a clear path of reasoning, and is not clogged by details.

Finally, are things organized in a way which will make it easy for the audience to remember, and have you considered the well known effects of memory when planning? The collection of the data, the facts, figures, and general information to be reported in the presentation, is only the first stage of preparation. Tailoring this material to achieve the aim of the talk, to interest and satisfy the audience, is an equal task. No one expects perfection, of course, and few of the points I have made in this chapter have easy answers. What matters is that you have thought about these issues. As I said at the beginning of this book, the average standard of spoken presentation is poor. And the main reason for this is ignorance about what needs to be done, and lack of preparation.

Notes to chapter three

1. Erskine, C.A., and O'Morchoe, C.C.C., Research on Teaching Methods: Its Significance for the Curriculum, *Lancet*, Vol.1 (1966), pp.709–11.
2. See also Walker, Stephen, *Learning Theory and Behaviour Modification* (Methuen, 1984).
3. Umberto Eco in J.Corner and J.Hawthorn (eds.), *Communication Studies: An Introductory Reader* (Arnold, 1980), p.145–7.
4. Donald Bligh, *What's the Use of Lecturers?* (Penguin, 1971), p.110.
5. Quoted by Ruth Beard, *Teaching and Learning in Higher Education*, (Penguin, 1976), p.107.
6. See Lindsay, Peter H. and Norman, Donald A., *Human Information Processing: an Introduction to Psychology* (Academic Press, 1977).
7. Max Wertheimer, *Productive Thinking* (Tavistock Press, 1961).
8. Peter Russell, *The Brain Book* (Routledge and Kegan Paul, 1980), p.125.
9. Erwin Bettinghaus, *Persuasive Communication* (Holt, Rinehart and Winston, 1980), p.159.
10. George Miller, *The Psychology of Communication: Seven Essays* (Basic Books, 1975), p.16, 31, 49.
11. Peter Russell, *The Brain Book* (Routledge and Kegan Paul, 1980), p.93.
12. Peter Russell, *The Brain Book* (Routledge and Kegan Paul, 1980), p.103.
13. J. Hartley and A. Cameron, Some Observations on the Efficiency of Lecturing, *Educational Review*, Vol.20, no.1 (1967), pp.30–37.
14. C.I. Hovland and W. Mandell, An Experimental Comparison on Conclusion Drawing by the Communicator and the Audience,

Journal of Abnormal and Social Psychology, Vol.47 (1952), pp.581–8.

15. Hermann Ebbinghaus, *Memory* trans. by D.H. Ruyer and C.E. Bussenius (New York, 1913).

16. Donald Bligh, *What's the Use of Lecturers* (Penguin, 1971), p.59.

17. Hans and Michael Eysenck, *Mindwatching* (Michael Joseph, 1981), p.146.

18. Faber, R.N., How we Remember what we See, *Scientific American* (May, 1970), p.105.

19. See also Arnold, Magda B., *Memory and the Brain* (Lawrence Erlbaum Associates, 1984).

Further reading

You will also find useful:

Baddeley, Alan., *Your Memory: a User's Guide* (Sidgwick and Jackson. 1982).

Miller, George A., and Philip Johnson-Laird, *Language and Perception* (Cambridge University Press, 1976).

Rodman, George R., *Public Speaking: an Introduction to Message Preparation* (Holt, Rinehart and Winston, 1981).

SUMMARY SHEET

Chapter three – Preparation

Good speakers are prepared.

The aim, not the subject, is what matters.

Prepare *more* material than you need.

Then decide what to leave out.

Keep down the unloading rate by:
— repetition
— more examples
— plunging in, and going back to explain
— mixing new information with familiar information.

Organization gives shape and pattern to a talk.

Human beings see patterns in these ways:
— Gestalt groupings
— by using our strong sense of physical location
— by making about seven divisons (plus or minus about two)
— by chunking complex details into groups.

Build the structure dynamically in front of them.

Construct a visibly logical argument.

Select typical examples: complete details are for writing.

Repetition and clarity replaces short-term memory by long-term memory.

'Recency', 'primacy', and regular rests help memory.

A well prepared talk can be outstandingly memorable.

4

Starting, carrying on, and ending

The opening

You have now reached that long dreaded moment – you have to stand up and start speaking. What do you say? How do you get started? Remember that the primacy effect will ensure that what you say in the first few sentences will be among the best remembered parts of the talk, remember that first impressions are lasting impressions: so how are you going to get started effectively? As with everything in speaking, thought makes the problem easier. There are a variety of starting tactics, and you can select between them on the basis of your own experience, as well as the type of audience. The opening which every speaker wants to make, but few succeed in pulling off, is the dramatic start with an arresting fact, quotation, or remark. Something surprising, exciting, disturbing, or plain unusual; something which will make the audience gasp with admiration, and sit up to take notice for the rest of the talk.

Not many speakers manage this effect, though everyone seems to dream of it. If you have such a fact or idea, it is certainly worth trying, for it does have the effect of alerting the listeners, and focusing their minds on the subject of the presentation. But it is difficult to achieve just the right tone of confidence, and drama, in the first sentences. One problem is that it is often difficult to get the first few words in the right tone, volume, and steadiness. Only when the voice has warmed up, can it be relied upon to produce the right effect. It is often more sensible, particularly if you are inexperienced as a speaker, to start with the simple matters of fact that the audience need to know.

What questions will be in the audience's minds at the beginning of the talk? To being with, simple practical matters like what is the talk about? Who are you? What are your qualifications, experience, and interests? How will it help them? Why should it interest them? What right have you to be speaking to them on this topic? These are the questions likely to be occupying the audience's minds, and they will need an answer before they will open their minds to the information you have to give. They will not make their memories available to you until they are sure the effort will not be wasted. If you do not satisfy some of these points, the lingering doubt will corrupt the input of information, and continue to interfere with their perception of your message.

Satisfying these questions is so important that even if you do start with a successful attention jerker, you will need to indicate answers to most of these questions within the first few minutes. You may have been introduced by a chairperson, who should have covered these points. But if he or she hasn't, try to fill in the missing details. It is undoubtedly much easier to listen to someone if you know exactly who and what they are, and what they are talking about. The easiest opening tactic is to reinforce what the chairperson said in introducing you. Extend it, fill in the gaps, but do not just repeat it. It gets you off the ground with the talk, and once you have started, it is easier to carry on talking.

The first task, then, is to establish rapport with the audience, gaining its confidence, and thereby making it prepared to give attention. To do this, explain how and why you are there, and what previous contacts you have had with this and similar organizations. It is also wise to check that you can be heard. These preparatory stages should not be allowed to take a lot of time; but briefly and clearly stated they are useful opening tactics. Thus, for example, a speaker's first sentences might be:

> I've been invited by Dr. White to talk to you about the software of voice-recognition programs. I worked on this problem for nearly ten years with JCN, a company very similar to yours. I now run my own software house, and have talked to many groups like yours. Incidentally, can you hear me all right at the back? I'm going to talk for about 30 minutes on three main topics . . .

To ensure that these important points are not missed, construct your opening sentences from a check list, such as this. Don't launch into an autobiography, each point needs only a single phrase, but it is useful for the audience to know these things:

1. Who invited me here, or arranged the talk?
2. What is the title of the talk?
3. Have I given a presentation to this, or any similar organization before?
4. What is my present job, or status, in which organization?
5. Can they hear me at the back?
6. How long am I going to talk for?
7. What are the main sections in my talk?

If you draft simple answers to these questions, and mention them in the first minute of your talk, you will help to ensure that the audience is content to listen to you.

Getting attention

After making clear who and what you are, you must launch into your subject without delay. Don't beat about the bush for several minutes, get into the meat of the subject straight away. One way of starting the talk is to put a question in the audiences' minds. Do they know how the raw material for the process is prepared? Have they thought about whether the I/O routines can be speeded up? Do they realize the financial drain on profitability which spoilage causes? Such a tactic focuses their attention on the issue, and helps them to listen positively to the information which follows it.

Another way of directing attention to a problem is an arresting quotation from a dissatisfied customer. Another way might be a photograph of a structure which has collapsed. It is also useful to point out how the present talk fits into previous talks, and a question related to the last presentation will help to remind the audience of what they already know, and how this new presentation will fit in. All these tactics have one central aim – to make sure the listeners realize what the purpose of the talk is, so they can fit the new information they are being given during the talk into a familiar conceptual pattern. Often speakers ignore this need to bring the subject into sharp focus at the beginning of the talk. In many presentations the consequences of neglect of the ideas and information are described at the end, rather than the beginning. In one lecture on dietary control, for instance, slides of the deformities which resulted from malnutrition were shown at the end; they would have been better shown at the beginning, so the audience could visualize the problems to be solved. 'These show what can happen; what can we do to prevent it?' would have made an excellent opening to the talk.

Asking questions is the best way to promote thought. Such questions may be only rhetorical, and not expect an answer from the audience, but Sime and Boyce showed that rhetorical questions raised the level of attention, and improved the amount of learning.[1] We are so conditioned to provide answers to sentences in question form, that our minds are subconsciously aroused towards an answer, even if we remain silent. Asking questions is an effective way of introducing a topic.

Other methods may also be used to increase interest and arousal. Advertisers typically use irrelevant messages about sex, status and emotions before selling their product. In the same way a stimulating fact or picture will arouse the audience and improve their reception of a quite different message which may follow. Remember Hillaire Belloc's aphorism: 'Tell them what you're going to tell them, then tell them, then tell them that you've told them.' The job of the introductory sentences is to arouse interest in what you are going to tell them, by telling them. Then the talk can go on to expand the subject, assured of attention from the listeners.

The need to arouse and prepare the audience is confirmed by psychological research. Many experiments show that unless the receiver is guided in how to decode the message, he may perceive something different. Psychologists have shown that knowledge about what a person is going to hear can change what he thinks he does hear:

> The English psychologist David Bruce recorded a set of ordinary sentences and played them in the presence of noise so intense that the voice was just audible, but not intelligible. He told his listeners that these were sentences on some general topic – sports, say – and asked them to repeat what they heard. He then told them that they would hear more sentences on a different topic, which they were also to repeat. This was done several times. Each time the listeners repeated sentences appropriate to the topic announced in advance. When at the end of the experiment Bruce told them that they had heard the same recording every time – all he had changed was the topic they were given – most listeners were unable to believe it. With an advance hypothesis about what the message will be we can tune our perceptual system to favour certain impressions and reject others.[2]

We may think that the experiment was unfair on the listeners; perhaps they were only trying to please when they invented sentences! But the central fact remains; we hear what we expect to hear. Therefore, if what is going to be said in a talk is announced at the beginning, the listeners more easily receive the message. Another experiment supports the same conclusion. Psychologists measured the effect of mental 'set' in perception by asking their subjects to repeat words which were flashed quickly in front of them, but after they were given different expectations about what they were going to see. In a typical experiment people were briefly shown the name of an animal, such as 'horse'. One group were told they would see the name of an animal, another that they would see the name of a flower, and the third only that they would see a word. People who were expecting to see the name of an animal recognized the word most quickly and made fewest mistakes. People who were not expecting to see any particular word did second best. And those who were expecting to see the name of a flower made most mistakes when shown the name of an animal. They also reacted more slowly.[3]

Listening to a complicated explanation, or a mass of unfamiliar facts, is similar to seeing words flashed too briefly in front of your eyes, or listening to a voice over a harsh mash of noise. They are all situations where the message must be disentangled from distractions. What the audience is told about the subject of the talk will condition what they understand the talk to be about. This is why it is so important to arouse interest in the subject, and be clear about what the purpose and content of the talk is going to be, in the first few minutes of the talk.

It is surprising how easily people are misled by what they expect to see, rather than what they actually do see. Abercrombie uses the following example. Read these labels quickly:

PARIS	ONCE	BIRD
IN THE	IN A	IN THE
THE SPRING	A LIFETIME	THE HAND

Only when asked to look more carefully do most people notice that the 'a' and 'the' are repeated in the middle of the phrase. We don't expect to see it, and therefore we don't see it. But if we are told, we see it easily. Telling people what they are about to perceive will radically affect what they do perceive.[4] The conclusion for the speaker is clear. Telling your audience in advance what to expect is an essential part of presenting information to them. In the face of such clear evidence, it is inexcusable to omit the preparation and warning phase of the talk. The subject must be made clear in the opening moments of the presentation.

Signposting all the way

The idea of 'signposting' originated with Tolman in 1951. The idea was that people become mentally disorientated by new information, and need to find their bearings. On an intellectual journey, signposts which point the way, and help to locate ideas, help people to understand. Tolman also speaks of a 'placing-need' which makes people want to have a map in their minds into which they can place the new information. Within that perceptual field clear orientating references and signposts are needed if the listener is to absorb information comfortably.

Once the opening stages of the talk are over, and the audience have been told where they are going, it is important to continue to signpost throughout the talk. This is done by announcing the topic, giving a heading, or listing keywords every time you start a new section of the talk. These can be written up on a board, flip chart or overhead projector. You should also give one or two sentences at the beginning of each section which act as an overview of the section. After developing the section, explaining and clarifying the point, giving examples, and discussing them, you should then come back to a sentence or two of summary and conclusion. Some signal is then needed to alert the audience to the fact that a new topic is about to start. Writing the new heading up, which requires you to change position, and pause while writing, is undoubtedly the best technique. But shifting position, allowing a significant pause, or even a change in the tone of voice is better than nothing.

The next section of the talk should start in the same way, with a sentence or two of definition, followed by explanation, examples, and clarification. As the talk progresses, you should also stop and take stock frequently, collecting together what has been said so far, summarizing the overall plan of the talk, and showing how what has been said so far leads on to the next point. Make cumulative summaries as you go through the talk. Each time you change topic and move onto a new subject, summarize in a sentence or two what you have said so far, refer to the map of the structure of the talk, and then announce the new heading. It seems easy and obvious, but many speakers do these things so quietly that no-one notices. The audience wake up from a day-dream to discover that the topic has changed while they were away.

So clear and repeated signposting is needed, if the talk is to be effective. Within each section, you should give the general picture at the beginning, and not launch into the body of the topic until you have given them an overview both of the topic itself, and of the way you are going to treat it. The effect of this is that within the overall structure each sub-element should have its own structure. Donald Bligh suggests that each point should be a version of the 'general form' of 'making a point' He lists these moves as follows:

> Concise statement
> Use the Board
> Re-expression
> Elaboration
> (a) More detail
> (b) Illustration
> (c) Explanations
> (d) Relate to other points
> (e) Examples
> Feedback
> Recapitulation and restatement[5]

A structure of this kind within each section will help to make the progress of the talk easier to understand and clearer. One of the difficulties that a speaker faces is that there is no lay out code in speech, such as the indentations and blank spaces which are used in written material to make the structure clear. The speaker must supply all these props to understanding with his voice. This is why it is especially important to emphasize the change of topic using as many different techniques as possible. Imagine a book in which the chapter headings were all set in the same size type as the rest of the page, and had no white space around them. If there were no paragraph breaks either, the text would be impossibly difficult to read. There would be a dense blur of information, with no visible shape or structure. Yet this is what happens in most talks. The paragraph breaks, and the white space round the headings must be provided by the speaker's tone of voice. Even if he emphasizes the change of topic, some of the audience may be day-dreaming at that moment, and miss the change. But if you write the new topic on the board, or flip chart, then when listeners return from their intermittent day-dreams they can see that a new topic has started. It is like leaving a message for an absent person to collect when he returns.

The absence of a layout code also means that listeners can't scan the page to see the shape of the information, or to look up a point which has gone by. Listening, unlike reading, gives the audience no opportunity to pause, rest and go back over material, at will. Once spoken, the information has gone. So the speaker has a much greater need for clear and simple structure in his information than the writer. The speaker must also be careful not to make mistakes; they can never be unsaid. And the listener must recognize that information lost is never recovered.

There are two more rules which must be added to the overall advice on how to glue together your points to make them clear, structuring a presentation. Firstly, there should be clear explaining links to connect point to point. Secondly, each individual point should follow the 'rule–example–rule' principle, where a brief statement of the fact, idea, or point, is followed by an example or illustration, and then that fact, idea or point is repeated. A simple phrase, at most a sentence, will do for the first statement. Any amount of illustration can reinforce this, depending on the importance and complexity of the point, and a summary re-statement should follow.

One final piece of advice; it is often very helpful to be quite open and honest about things you find difficult to explain. By taking your listeners into your confidence, you will enlist their interest in the solution to the problem of how to explain the point. You will also make them feel that their difficulty in understanding is not because you are a bad explainer, but because the point itself is complicated. You align yourself with them, and make the point itself the enemy. They are then more sympathetic, more aware, and in trying to help you, will accelerate their own understanding.

The longer speech

A longer presentation (one which lasts more than ten minutes) demands a long span of uninterrupted attention from the audience, and therefore needs more skill in the structuring of the talk. Ten minutes may seem a short period, but as I have said earlier, audiences find it difficult to listen for long without taking little breaks for day-dreams. So longer talks need more organizing, more linking, and more reminders. They have a greater overhead of time which must be devoted to housekeeping activities, like keeping tabs on where the talk has got to, and keeping the structure fresh in the audience's mind.

Signposting in a longer talk becomes more important. The speaker must provide a thread to help the audience to find their way through the maze. He or she must remember how limited any listener's span of attention is, and offer regular directions for the lost travellers. The basic rule is that the receiver of the message always needs *more* explanation than the speaker thinks he does, because he is not as familiar with the material.

In a longer talk some technique must be found to interrupt the monologue with a different activity every five or ten minutes. Break up long stretches of time by strategically timed visual aids. An alternative is to schedule occasional brief periods of discussion to provide relief. The audience are stretching their legs mentally, if they are given a chance to talk themselves. The discussion period doesn't need to be long; a few minutes relaxed talking as a group will help the listeners approach the next section of the talk feeling refreshed. If the speaker breaks up a long talk, and provides variety in this way, the audience will experience it as several short talks rather than one long one.

All this advice on how to reduce the burden of listening by breaking up a long period into shorter ones is not based on laziness. Psychological research gives clear evidence that shorter sessions improve learning. The early work on memory showed the importance of rest periods. Hermann Ebbinghaus, for example showed that the efficiency of learning improved when he included short periods of rest between learning sessions. At first he was surprised by this, since he expected periods of rest to cause people to forget some of what they had just learned, and so reduce the overall amount of learning. But he realized that the reminiscence effect was causing learning to improve. The conclusion from the experiments was that both primacy and recency increase the efficiency of a learning session which is punctuated by breaks. A single session benefits from primacy and recency only at the beginning and the end. But if the learning task is broken into several shorter sessions, with breaks in between, there are more occasions when the primacy and recency effects can assist learning.

It seems that the memory, like some muscles, tires easily, but recovers quickly. Ebbinghaus's results showed quite clearly that the benefit of a break increased as the length of the break increased up to a maximum of ten minutes. After that, lengthening the break to a quarter or half an hour made no difference. The result has been confirmed by many subsequent researchers. This is why most effective courses and conferences schedule ten minute breaks every hour or so. It is also why a presentation which lasts more than ten minutes needs to have built in variety, and breaks of various kinds, so that the listeners can recover their mental energy. The key to an effective talk is variety, whether the talk is long or short. And the key to an effective longer talk is to break it down by whatever means available into a sequence of shorter sessions. If you spend time devising ways of breaking up the long session, you will be rewarded by an alert and attentive audience.

The closing stages

Having started successfully, and carried the talk on effectively without losing the audience's attention, time is up and you now have to finish. How do you do this successfully? There are tactics for finishing, just as there are tactics for opening, and thought about what you are trying to achieve will, as always, improve the performance. Many people feel that the ending is more than half the battle. Certainly, the impression which the audience will carry away with them will be strongly influenced by what happens in the last few minutes of the talk.

The essential aim is to round off the presentation on an up beat. You can, for instance, get attention again by a vital, arresting and memorable fact or idea. Another way of finishing is to tie up all the loose ends by restating the sub-headings you used, restating the main heading or title of the talk, and restating the conclusion you came to. But whichever tactic you choose, it is important to remember that the last sentences must be telling. So the encoding you chose for your closing remarks should be memorable. Try to find a good phrase, a witty or stylish way of putting the point, or some clear statement of the main aim of the talk, for the last thing you say. It can help to have the last sentence or two written down in your notes. If you are nervous about forgetting it, or getting confused, it may even be worth trying to learn it off by heart.

The virtue of all these tactics is that they will save you spoiling the effect of the presentation by falling into a weak or confused ending, which trails off in embarrassment. A surprising number of speakers seem unable to end firmly, but mumble on with increasing indecision at the end of their talk. Never end weakly with: 'Shall I go on? . . .'; or 'What I should have said if I'd had time was . . .'; or 'What I intended to say was . . .'; or 'I think that's all I have to say'. The audience will remember the last point, or sentence, clearly. If that last sentence is a shambolic confusion of indecision, with the texture of a rice pudding, then the whole talk will be remembered as weak. End boldly, with a final statement of your main point which you fly like a banner, before sitting down.

The aim of the concluding sentences is to make sure that your talk goes somewhere. It should not just peter out in confusion. Karl Lashley told a nice anecdote:

> I attended the dedication, three weeks ago, of a bridge at Dyea, Alaska. The road to the bridge for nine miles was blasted along a series of cliffs. It led to a magnificent steel bridge, permanent and apparently indestructable. After the dedication ceremonies I walked across the bridge and was confronted with an impenetrable forest of shrubs and underbush, through which only a couple of trails of bears led to indeterminate places.[6]

Make sure that your proudly constructed talk does not lead to a wilderness of bear-tracks! It is also a courtesy, if you are speaking as part of a longer seminar, conference, or presentation, to prepare the ground for the next topic and speaker. Something simple like: "It's now coffee time. After a ten minute break, Alan will tell you about the stress calculations used in the project," will form a neat conclusion. This tactic helps to give the audience a sense of continuity.

If you are not followed by someone else, make sure that you end as strongly as possible. 'So we see that nutrition is a vital element in the health of the community', or 'Voice-recognition is developing rapidly, and within ten years will be commonplace', or 'the familiar chlorate process, which is the mainstay of our company profits, is much more complex than most of us realize', is the sort of clear statement that is needed. If you start clearly, keep people aware of where you are going throughout the talk, whether it is short or long, and end firmly and impressively, your talk is going to be remembered as an effective presentation. Judging by the average standards of presentation one hears, it may well be the best of the day.

Notes to chapter four

1. Donald Bligh, *What's the Use of Lectures?* (Penguin, 1971), p.165–6.
2. George Miller, *The Psychology of Communication: Seven Essays* (Basic Books, 1975), p.79.
3. W.R. Garnier, *Uncertainty and Structure as Psychological Concepts* (Wiley, 1972).

4. M.L.J. Abercrombie, *The Anatomy of Judgement* (Penguin, 1979), p.31.
5. Donald Bligh, *What's the Use of Lectures?*, p.103.
6. John P. de Cecco (ed.), *The Psychology of Language, Thought, and Instruction* (Holt, Rinehart and Winston, 1969), p.249.

Further reading

You will also find interesting:
Greene, Judith and Carolyn Hicks, *Basic Cognitive Processes* (Open University Guides to Psychology, 1984).
Henriques, Julian, Wendy Holloway, Cathy Urwin, Couze Venn, Valerie Walkerdine, *Changing the Subject: Psychology, Social Regulation, and Subjectivity* (Methuen, 1984).

SUMMARY SHEET

Chapter four – Starting

Dramatic openings are difficult to pull off.

Tell them who you are, where you come from, and what you know.

Then start into the subject with a good example.

Audiences need to be told what they are listening to.

People see what they expect to see, not what's there.

Continue signposting throughout the talk.

Each section of the talk should have a statement, examples, explanation, and then a summary,

Speaking has no layout code, like the white space round a title, or paragraphs, in printed text.

Longer talks need clearer structures, and more variety.

Break up one long talk into several shorter sessions.

Finish cleanly, with a clear statement of the main point.

5
Making notes

Script or notes?

One of the first problems a speaker faces when he or she starts to prepare his presentation in earnest, is how is he going to record it so that he is reminded as he talks? In other words, what sort of notes is he going to make? Nine times out of ten, this question is never considered. Some sort of notes are produced, usually depending on factors such as what other people have been seen to do, what sort of notes were used at school, and sheer chance. Notes to speak from seem just to happen, without thought, and the speaker muddles through. Poor notes, however, are an added strain when talking, and can cause you to miss sections of the talk, lose the place, and dry up. So it is worth thinking about the best way of taking notes. As with everything to do with speaking, a little thought in advance saves a deal of embarrassment and confusion on the day.

The first question is, should you write down the talk and read it out, or use some sort of note form? And should those notes be full and detailed, or skeleton notes? To help solve this first problem, let me describe four ways of preparing a talk, showing the advantages and disadvantages of each.

Written scripts

In many ways the most obvious thing to do, and often the fiirst method chosen by inexperienced and nervous speakers, is to write the talk out in full, and read it out from the script. This is a method which is often used for technical papers, and symposia, and is therefore one which a young professional is most likely to be familiar with. He or she may even think it is the *only* proper way to deliver a so called talk. In symposia the text may even be distributed in full before the talk, and time spent solemnly reading it out during the meeting. Anyone who has been to a conference where this is done, knows that it is quite impossible to listen to the text being read out. It is painfully monotonous, and has little to do with the experience of talking to someone who is an expert in the subject. The process is only called 'speaking' by courtesy, and because a voice is used. It has more in common with a public reading, usually being given by someone peculiarly inept at reading aloud.

Why, then, do people write full scripts, and read them out? Partly because it seems to be expected of them, partly because of the obvious advantage of a written script – it gives the speaker confidence. But the disadvantage is overwhelming: written language is not spoken language. This statement needs some clarification. Here is an example of spoken language – an accurate transcript of a BBC broadcast:

> Well if you take one of these animals and put it between two electrical terminals in a laboratory, and create a strong static electricity field, which doesn't hurt the animal at all, it's perfectly lively and unaffected by it, but it will start to discharge electrons; they fan out from the openings of the body, the openings in its external shell, its exoskeleton, and there's an avalanche of electrons moving out and knocking into molecules of gas in the air, nitrogen molecules mostly, and these are excited, and because they are excited, they glow, and so each individual insect gives out rather a weak light, but if you look at it, in a darkened room, you can see this glow fanning out in all directions.

As you can see, it is very different from written language. If the same conversation were written, it would presumably go something like this:

> Harmless laboratory experiments demonstrate that the animal dis-charges electrons in a strong field of static electricity. They fan out, avalanch likè, from the openings in the exoskeleton, exciting nitrogen molecules in the air which then glow. As a result the insects give out a weak light, which, in a darkened laboratory can be seen to fan out in every direction.

We have got so used to the kind of simulation of spoken language used in novels, that we fail to notice what people actually say when they are speaking. Spoken language, when it is carefully and accurately transcribed from a recording, has little in common with written language. Among the main differences are the fact that the grammar of spoken language usually seems strange, even obscure, when the passage is written down. This is because the *structure* of a spoken passage is made obvious by the intonation – the way the voice goes up and down in pitch, emphasizing certain words, and making pauses between phrases. Spoken language doesn't need grammar, to give a meaning to the sequence of words. Indeed, to put it another way, the grammar of written language has probably evolved in order to replace the intonation of the spoken voice. And this is what causes the problems, for just because written language has grammar, it doesn't need, and doesn't encourage, intonation. In other words, the written form positively encourages the use of a flat monotonous voice; indeed it may be difficult to use intonation and variety when reading written material, because the grammar will make it superfluous. The difference between spoken and written language can be summed up in one simple fact: spoken language doesn't have sentences. If you look carefully at the transcript of the BBC talk above, you will notice that it is not really possible to decide where one sentence ends and another starts, and certainly few of the sentences are grammatically complete. We speak in a collection of phrases, not in sentences.

Spoken language is not written language.
Spoken language has two other important differences from written language. Firstly, speaking uses much repetition. The typical way of explaining, amplifying, and exploring a point is to add an extra phrase in a sort of phonological bracket. By dropping or raising the voice, it is made clear to the listener that the information is a sort of sideline, or footnote, which is meant to clarify what is being said, rather than introduce a new point. Speakers also tend to restart sentences in different ways, trying to get across what they mean by different routes, and when they feel the point is clear, not bothering to complete or tidy up what they have started to say. There is also much trying out of different words, and rhetorical repetition for emphasis. All these features contribute to the muscular, flexible, and alert feeling of spoken language. It is like a living contact with the mind of the speaker, whereas written language is a fossil record of his or her thoughts.

The second way in which spoken language differs from written language is that the choice of vocabulary is very different. Written vocabulary is formal, and explicit. Spoken vocabulary tends to be familiar, and everyday. Indeed, it is usually possible to get someone to simplify and clarify a tortuous written sentence by asking him to look away from the page, and say what he means. A writer who has solemnly written: 'Tests were conducted on the loader to ascertain the maximum failure capacity'. when asked what he meant, would say something like: 'We loaded it up until the cable broke'; a simpler, and clearer, way of explaining a technical point.

Writers, then, use formal grammar, single expressions, and elaborate, abstract vocabulary: speakers use intonation, repeat things until they are clear, and use everyday words. There are great differences between spoken and written language, and when written language is read out, it is less effective. I am not suggesting that there is a difference of worth, between written and spoken language. They are simply used for different purposes; one is to communicate face to face, the other communicates remotely. Misusing the difference is one cause of boring presentations. It you read out written language, your voice will naturally lack intonation. The structure of what you say will be over formal, and the vocabulary will be too abstract. This is why listening to written papers being read out is so difficult. The listener gets no sense of contact with the speaker's mind – there seems to be a wall of fog between the living mind of the speaker, and the listener.

Written language often sounds false and clumsy when it is read out, and what I have said in the last few paragraphs should explain why. The added problem is that many people are poor readers; their reading voices are stumbling and monotonous. It *is* possible to read written text in an interesting way – actors do it constantly – but it requires great skill. It is certainly not to be recommended as a way of giving a technical or informative presentation.

Another disadvantage of reading is that the presenter loses eye-contact with the audience. Because he or she has to follow the text, it is impossible to do more than glance up at his listeners from time to time, whereas someone speaking spontaneously will naturally be looking round at the listeners. When reading, a presenter also loses the chance to make gestures and arm movements, which are naturally suppressed when reading from a script because they seem artificial. None-the-less reading a written text is a method often used. It is one I don't recommend. In all but exceptional circumstances, it is a sure way of losing the attention and interest of the audience. It is an expensive way of buying the confidence that you won't forget what you are going to say. You may not forget, but the audience almost certainly will.

If you are terrified of forgetting what to say, there is a compromise which helps boost confidence, by providing safety points to return to if the impromptu flow of words breaks down. The technique is to write down the opening and closing sentences, as well as sections within the speech, for use in the case of emergencies. By providing islands of security, you will increase your self-confidence. It also provides natural resting places, and if the worst happens, and you dry up, there is something to say while you are finding your feet again. But don't write down more than a few sentences, otherwise the whole talk will acquire the monotonous flavour of the written script. The first sentence of each new topic, and the conclusion of each section, is as far as you should go. Inbetween, use ordinary notes.

The learned text

A second method sometimes used by inexperienced speakers is to write the talk out in full and learn it by heart. The advantage is that it avoids the loss of eye-contact and gesture. But the method has disadvantages; there is the danger of forgetting the lines, and anxiety about remembering them is an added strain. The speaker is also at the mercy of interruptions, which can make him or her lose the thread of the talk. Another disadvantage is that the speaker spends time and energy in wasteful rote learning, and therefore has less time and energy to use in preparing the content of the talk. But the main disadvantage of the written text still remains. When written language is spoken, whether it is read or remembered, it sounds stilted, formal and unnatural.

From notes

The third method, and the one which all experienced speakers recommend, is to prepare the talk carefully, but to deliver it from notes, choosing the actual phrasing extempore. The great advantages of this method are these: firstly it has flexibility, spontaneity and openness. Feedback from the audience can vary the structure and content of the talk to fit in with their needs and interests. The second advantage is that there is eye-contact and natural gesture. The speaker can look at his audience while he talks, and his gestures are a spontaneous reflection of the structure of his thought. These important aspects of speaking are discussed more fully in chapter nine. Having only notes for most of the presentation forces the speaker to talk to the audience naturally and spontaneously. He or she will probably find this easier, and more enjoyable, than they expect.

All experienced speakers use this technique of talking from brief notes. Research confirms that it is the most acceptable method for the listeners. Coats and Smidchens (1966) show that when a lecture is constructed around a few key points it is much more effective for the listeners than hearing a paper being read.[1] The only exception to this rule is that it is useful to read quotations. It gives a variety of approach, which is the key to interesting (that is effective) speaking. It also lends an air of authority to the quotation, because the audience accords it the extra credibility allowed to the printed word.

Not prepared

A fourth method, worth mentioning for completeness, is speaking without any written preparation at all. As a way of preparing a talk it has no advantages; its disadvantages are that you may imagine you have prepared more in your mind than you actually have. Only writing it down will show what is really there, help to focus and clarify your thoughts, and identify the points you want to make. The other obvious disadvantage it that it makes it too easy to forget; you are likely either to dry up completely, or to miss out important points. It should never be used except for unexpected invitations to speak. Without notes, a talk usually lacks a plan and a structure, so speaking without notes should be confined to very short speeches. If you suddenly find that you are expected to talk, if possible, make scribbled notes on an envelope in the last few seconds while you are being introduced. If you also move slowly while the person who introduced you sits down, you stand up, and everyone shuffles about, it will give you added time to think. There is no need to worry about the pause; the audience will be glad to have a short break from listening.

There is only one rule for these unprepared, impromptu, talks. Make only one point, one joke, or tell one anecdote, then shut up. Brevity is the only virtue of the impromptu talk. The exception is the speech given by the experienced after-dinner speaker. Like an actor, he has a sheaf of jokes and anecdotes tucked away at the back of his mind, and has only to staple together two or three familiar routines before he talks. He can then run on ad-lib like a gramophone in a groove, apparently totally confident. It is an illusion. Unless you have the experience, don't try to emulate it. It's like trying to climb Mt Everest without the professional's climbing equipment. You will fall off!

Notes are to help you

The notes you make are the most important insurance policy for the success of the talk. The product of the preparation stages is a set of notes, and they represent the only permanent part of the talk. Speaking is ephemeral, while notes endure. But notes are not the whole talk. You will find, as you talk, that ideas and facts from the work you did in preparation will come back to you, and you may decide, impromptu, to use a piece of information which you did not put in the notes. There is nothing wrong in this: the purpose of a talk is to say what you know about a subject, and notes are for assistance, not to replace knowledge.

The main advantage of good notes is to ensure that you do not forget what you intend to say. A great deal of research has been done on memory (there are many specialist textbooks on memory, for instance), and one of the most consistent results is that stress affects memory. Its usual effect is to make us forget important things, but stress can also cause complete black-outs of memory, as well as causing sudden vivid remindings about things previously buried in the subconscious. The effect of stress is unpredictable: it makes memory irrational and random. And the speaker is under as much stress as most people experience in their day to day lives. It is therefore especially important that he or she takes steps to compensate for the erratic and unreliable performance of the memory under stress.

Some people are unfortunate enough to go completely numb and silent when facing an audience – their memory switches off. The brain processes that operate recall are notoriously out of reach of the will power. We are quite unaware of the process of laying down memories, we feel no pain, no sense of effort, and no sense of choice. We can only predict, in a fairly random way, what we will find memorable, and what we are likely to forget completely. Brain specialists believe that quite large parts of the grey matter are involved in the recording and recall of memories, just as we now know that huge parts of the brain are involved in decoding the information from our retinas, before passing it on to the conscious part of the brain. But we are not aware of the process of stereoscopic vision, just as we have no consciousness of the processes of memory. We often need some object to remind us; notes are a kind of external memory that is under conscious control. Notes jog your memory, and produce what the audience perceives as a fertile flow of ideas and enthusiasm.

Some people, instead of seizing up in front of an audience, become uncontrollably garrulous under strain. They always find plenty to say; the trouble is that it may, or may not, be relevant. Good notes are just as important for this kind of person. The art of good talking is not just to fill the alloted time; it is to use the time wisely to say as much as possible that is useful and necessary. The most useful function of notes is not just to remind you of the material, but to give it structure. They provide a plan or map of the structure of the talk.

Notes are the main way in which the content and structure of the talk can be controlled. Without notes, most talks are formless ramblings. With notes they can be an orderly set of points, with a clear sequence and coherence which the audience can rely on. Notes should not be thought of just as bits of information to fill the time. Notes are like pigeon holes, into which the subject can be fitted. But the notes are not the pigeons. The facts, ideas, information and anecdotes will come from the speaker's memory; he or she, after all, is the expert on the subject, and the talk will be more interesting if it is spontaneous and anecdotal. The notes provide the structure of categories, the wood round the pigeon holes, to continue the metaphor, which controls and shapes this flow of information, knowledge, and stories. For this reason, notes should have a prominent and logical sequence of headings. Because their main function is structural, they can also contain cues, quotations, jokes, signposts, and stage directions such as when to stand up, sit down, move to the board, and change to a new topic.

Notes, therefore, should not be a version of the full information. The details are much more interesting, and convincing, if they come directly from the speaker's memory. The speaker should be like someone engaged in earnest, animated conversation, anxious to tell his listeners about all the facts and ideas he has at his finger tips. If the notes are a dense maze of factual material, he will become more like someone saying his lessons. So notes should be the mere prompting, the skeleton, on which the talk can be built. All sorts of information can form these promptings; but they should consist of thoughts, key-words, and headings, not full sentences.

One way of preparing notes is to write the key words boldly on the left hand side, leaving the detailed notes on the right hand side. The purpose of this is to make the structure of the talk as clear as possible to the speaker – then he or she will in turn make it clear to the audience. Figure 5.1 shows an example of notes arranged in this way.

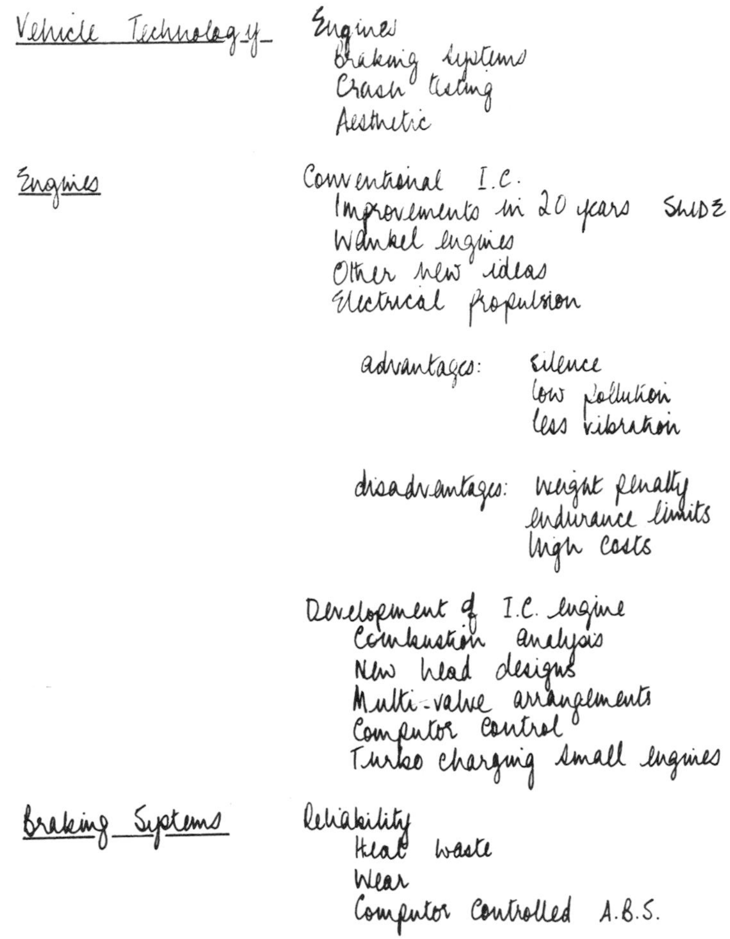

Vehicle Technology —   Engines
                         Braking systems
                         Crash testing
                         Aesthetic

Engines   Conventional I.C.
                         Improvements in 20 years   SLIDE 1
                         Wankel engines
                         Other new ideas
                         Electrical propulsion

                              advantages: silence
                                            low pollution
                                            less vibration

                              disadvantages: weight penalty
                                              endurance limits
                                              high costs

                         Development of I.C. engine
                         Combustion analysis
                         New head designs
                         Multi-valve arrangements
                         Computer control
                         Turbo charging small engines

Braking Systems   Reliability
                         Heat waste
                         Wear
                         Computer Controlled A.B.S.

Fig. 5.1 Well laid out notes

Colours and cards

Notes do not have to be written all in one colour, or in one size of
script. You should experiment with using coloured pencil or biro to
distinguish different elements in the notes. For example, some
speakers write quotations in red; it helps to locate them quickly when
they want to anchor the talk back to the planned sequence. One
system is to write central points in black, and less important or
peripheral material in green. Notes in green are only used if time is

dragging, and the material is needed to fill out the talk. If time runs out before the speaker has got through the main material, then everything in green can be skipped over. It also helps to have different sizes of script for different levels of importance. Size and layout should always be used to make the key words stand out.

One of the common failings in the preparation of notes, is to write them in normal handwriting. As the material gets denser, so does the handwriting, and you find yourself making late additions to the notes by squeezing in tiny, cryptic, messages to yourself. Avoid making tiny marks in the notes, which may be visible when sitting at a desk, but are invisible (or, worse, can be misread) from the much greater distance between eye and paper when standing up. Few things sap the confidence of an audience more than seeing the speaker peering myopically, and in obvious confusion, at his notes. Do make sure that your notes are large enough to read comfortably. Headings, and main key-words should be in double size capitals.

The best materials for notes on are 8" × 5" cards. Use one side only; it makes them easier to refer to, both while preparing the talk, and in the heat of the moment. Cards have a number of advantages. Thinking about the audience first, they give the audience hope and confidence when they see a rapidly diminishing pile of thick cards, rather than a sheaf of large, thin paper, which never seems to get any smaller. Another advantage of card is its stiffness. Cards are much easier for nervous fingers to hold; paper is flexible, and acts like a sounding board for every shake and tremble of the hands. Card does not shake. A final advantage is that cards can easily be shuffled into a new order, and the ones which are not needed discarded. John Mitchell endorses this advice, in common with most experienced speakers:

> The use of cards gives the impression that the speaker has organized his subject and knows so much about it that he needs only to be reminded what to discuss next. These cards hold considerable information, and they do not terrify an audience as much as a great sheaf of paper does. Further, they can be managed with one hand and do not rustle in a microphone. Finally, they give the audience a visible cue whenever the speaker turns to a new topic.[2]

There is no doubt, then, that the best technique for notes is to write them on cards. Use one major heading per card: the heading can be on the top left corner, in bold letters, and the notes spread out with key-words on the left, and facts, figures, individual points, and quotations listed on the right. These cards can be shuffled to change the order of the points, either when you find yourself running out of time, or when you have to prepare a new talk on the same subject. Cards also make it easy to add new material, since a whole new card can be prepared, and slipped in between the others. Cards have two final, decisive, advantages. They are easy to carry around. They can be fitted into a pocket, so if you want to revise your talk on the train, or even in the washroom just before making the presentation, you can keep them in your coat pocket. Secondly, card is easier to hold, and you won't find yourself fumbling with trembling sheets of paper which show an unconquerable attraction for the floor, and blow about in breezes from any open window. In fact, cards have so many advantages, that it is difficult to understand why anyone uses anything else. The answer, I suppose, as with so much else in the technique of speaking, is that most people have simply never thought about it. Or rather, they have never been told about these simple facts and techniques. The best size of card to use, incidentally, is an 8″ × 5″ card , the common 5″ × 3″ are far too small to see when standing up. Any stationer will have the larger size of card. Figure 5.2 shows a set of cards, prepared for a talk.

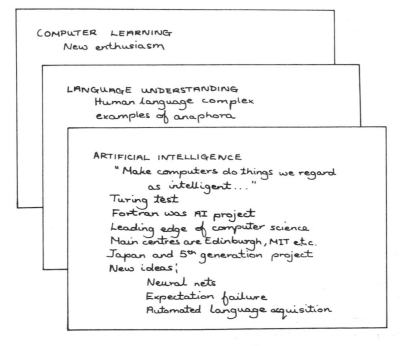

Fig. 5.2 Notes on 8″ × 5″ cards

Finally, always read through your notes ten to fifteen minutes before giving the talk. Then the key points and overall structure of the talk will be clear in your mind, and a simple glance at the notes will remind you of the point to be made. The content of the talk will then be secure, and you can concentrate on the art of presentation during the talk itself.

The audience can see them

Speakers sometimes seem to be trying to hide their notes from the audience, as if notes were something disgraceful, not to be seen. This is silly: there is no shame in using notes. There is no point in trying to pretend that the whole thing is off-the-cuff: the audience will have much more confidence in what you are saying if they can see that you have made thorough preparation, and are speaking according to a carefully thought out plan. But do remember that when you pick up your notes they are directly between your eyes and the audience's. To the listeners sitting in front of you these notes are exactly at eye level. Especially if the listeners are bored (and what audience isn't sometimes a little bored) the notes will provide something to watch. And watch they will. So be very careful what your notes look like and what you do with them.

A lot of nervous speakers fiddle with their notes, fold and unfold them, or flap them like a fish out of water, or a torn sail in a gale. I even saw one speaker methodically tearing his notes up, one by one, as he finished the page. The audience, who had the feeling they were watching a horror film – with a mixture of delight and fear – were all waiting for him to find that he had torn up the wrong page. Calm and controlled handling of notes gives the audience the right signals about your state of mind. Do not shuffle your notes like a Las Vegas gambler, and above all, try not to drop them.

What about their notes?

Having thought about your own notes, you should also think about the audience's notes. If you are giving a talk which has facts and information in it, the audience will probably make their own notes. Consider how they will transfer your notes into their own. Note-taking is always associated with college education, but many of your audience, if they are professional people, may well be graduates and have acquired the habit of note taking. They will make jottings during your presentation.

These notes are not beyond the control of the speaker; there is much he or she can do to shape what the audience writes down. The listeners who are making notes will, for instance, write down anything which you emphasize. If you make the heading and key-words clear, they will write those down. They are also likely to write down any figures you give, as long as you give them slowly enough, or write them on a board or flip chart. If you structure the talk clearly, and write down key-words and figures, the audience's notes, as well as their understanding and memory of the talk, will be clear. In colleges, the popular lecturers, even with the most cotton-headed gigglers, are always those who provide a clear structure for the student's notes. Research shows that making notes helps people to remember. Michael Howe studied the way students use notes, and showed that things which were noted were six times more likely to be remembered than things which were not.[3]

Hartley and Cameron (1967) recorded the items in a presentation, and checked to see how many of them appeared in notes taken by the audience. Rather less than a third were recorded, and only half of what the speaker had stressed as important. But most things which had been written on the black-board were recorded, showing that such highlighting of key words and concepts is a valuable way of making them memorable.[4]

In trying to ensure that the audience's notes are good, the most fatal error is to write them yourself. You may think that if you prepare a set of notes which accurately summarizes the points you are making, and distribute them to the audience, you have solved their note taking problems. Not so. Freyberg (1956) showed that where this was done the listeners took no notes of their own, and remembered far less than when this was not done.[5] It may seem helpful and efficient to prepare notes for the audience, but it turns out to be counter-productive. If you distribute these notes before the talk, the audience will not bother to listen to the talk, because they think they already have the material in note form. If you distribute the notes after the talk, those who have made their own notes will be angry that they have wasted their time. The rest won't read them anyway. Either way, distributing your own notes is a poor technique.

The best solution is to prepare a single sheet of paper with the five to seven headings you are going to use on it, with blank space between these headings for the listeners' own notes. This will help them to see the shape of the talk, and watch progress from point to point, to see how they are getting through the journey. If will also help to focus their attention on the key-words, while ensuring that the activity of note taking reinforces their memory. Psychologists proved some time ago that the activity of writing notes increases many times the efficiency of memory. Presumably because the material must pass through the mind several times – when heard, when summarized, and when written down – thus reinforcing memory by repetition. Active participation, rather than sitting letting the facts wash over, also helps the listeners to remember. Note taking is undoubtedly the best way to remember things. Even so, don't assume that every word you say will be remembered. Hartley and Cameron (1967) showed that only between 21% and 73% at best of what the speaker considered important was preserved in students' notes of a lecture. Expect a large wastage rate on your information, and stress and repeat those points which are important.

So the conclusion of research is that the best way for an audience to make notes is to write down the keywords the speaker has used. Michael Howe's work, referred to above, found that the higher the percentage of key words present in the notes the better was the recall.

So you will help your audience most if you stress the key-words in your talk. By emphasizing that these words are the linch pins of the talk, and writing them down on a board or flip chart if one is available, you will underline their importance. The best presented talks should result if a dozen or so key-words are firmly planted both in the audience's notes, and in their memories.

The audience's notes, then, are as important as your own notes. By keeping the audience's attention you can avoid falling foul of the well-known students' definition of a lecture, as a period when the lecturer's notes are transfered to the students notes without going through the mind of either.

Notes to chapter five

1. Coats, W.D., and Smidchens, U., Audience Recall as a Function of Speaker Dynamism, *Journal of Educational Psychology*, Vol.57, No.4 (1966), pp.189–91.
2. John Mitchell, *A Handbook of Technical Communication* (Wadsworth, 1962), p.187.
3. Howe, M.J.A., and J. Godfrey, *Student Note Taking as an Aid to Learning* (Exeter University Teaching Services, 1977).
4. J. Hartley and A. Cameron, Some Observations on the Efficiency of Lecturing, *Educational Review*, Vol.20 (1967), pp.30–37.
5. P.S. Freyberg, The Effectiveness of Note Taking, *Education for Teaching* (February 1956), pp. 17–24.

SUMMARY SHEET

Chapter five – Notes

Speakers often don't think about notes.

Inexperienced speakers write a full script and read it.

Spoken language is different from written language.

Speaking written language reduces intonation and interest.

Rehearsing a learned text is risky, and tedious.

Experienced speakers always use notes.

Unprepared talks are only for the experienced actor.

Notes support the memory.

Notes make the structure clear to both speaker and audience.

Use different colours for headings, optional material, and the main notes.

Use 8″ × 5″ cards. They are easy to modify, easy to hold, easy to carry, and look better.

There is nothing to be ashamed of about using notes.

Don't write the audience's notes for them.

They will note figures, headings, anything you write up, and what you emphasize.

6

Coping with nerves: the credibility problem

A common complaint

Although it has been left until half way through the book, nervousness is probably the biggest problem to be surmounted for most in-experienced speakers. Were it not for nervousness, common sense, and normal intelligence, would ensure that most talks were interesting and well planned. But nervousness seems to disable common sense, and normal intelligence gets swamped by anxiety. A book like this is needed just because speakers get nervous. Like a rabbit caught in a car's headlights, they don't know which way to run. All sorts of bizarre behaviour results, unless there are firm guide lines. Like clinging to the wreckage in a storm, any fragment of advice gives security. Even if the speaker doesn't feel at all like smiling, for example, the knowledge that he or she ought to smile is enough to make them feel that they are doing the right thing.

Nervousness is a very real problem, and is the root of most of the other problems with speaking. We all talk competently in a group of friends, but as soon as the group of friends becomes a wall of strangers, nervousness usurps our every-day competence, and we need the prop of advice.

Unfortunately, many books on speaking dismiss nervousness as not worth discussing. Like laziness, or cowardice, these books seem to imply that it is something to be ashamed of, and certainly not something to be discussed. The speaker may be jollied along with advice like 'Don't worry', or 'It'll be all right'. He or she is given the impression that nervousness, like incontinence, is something which is better not thought about. It will go away if you ignore it, and if not, there's nothing to be done about it. Nervousness is beyond help, these books seem to imply, and only courage will overcome it. Good chaps put a brave face on it, and never mention it to other chaps.

Sadly, all this hearty pretence is no help; it merely increases the sufferer's sense of his or her own inadequacy. It is also cruel: extreme nervousness is one of the most unpleasant experiences most civilized people go through. It is a form of physical and mental suffering which is unparalleled. Extreme misery, anguished anxiety, and even physical nausea are added to shame and a sense of inadequacy. Embarrassment is the least of the suffering. It may take weeks, months even to get over the misery caused by a catastrophic failure to cope with nervousness. The speaker may go through savage reassessments of his or her abilities as a result of ruining a presentation through nerves. Undoubtedly, nervousness is a serious problem; it needs careful and considered help.

Nervousness *can* be helped, and eventually reduced to manageable proportions. It is, after all, a purely mental phenomenon. Attitudes, and knowledge about the cause and function of the anxiety, advice about how to reduce it, and experience which renders the terrifying familiar, are the clues. Much of the work on nervousness has been done by musicians: talented young musicians find the intricate dexterity required to play their instruments turned into clumsiness in front of judges and audiences. Since it is clearly a waste, musicians have studied the problems of tension in performance. Speakers, who have similar problems, can benefit from the knowledge and techniques gained from these studies.

The first thing to learn about nervousness is that it is universal. Every nervous speaker thinks that he or she is the only one in the world to suffer. Compared with the calm competence of every one else, he or she feels their own shameful failure as a personal inadequacy. The truth is that nervousness when facing an audience is very common. Almost everyone suffers from nerves, even experienced professionals, and the reason why we are not aware of this is simply that the basic effect of nerves doesn't show. Providing the gestures are controlled, butterflies in the stomach are invisible to the audience. So the calm and confident speaker you watched with envy, was almost certainly trembling like a leaf inside: you just couldn't see it.

It is a good thing that speakers are nervous. Contrary to popular belief, the calm and controlled speaker is acting, he or she is disguising nervousness in a practised simulation of indifference. If he or she were really not nervous, there would be no energy to give the talk: nerves are useful to the speaker, without them he would go to sleep. Even people who make their living from appearing in front of audiences – actors, comedians, performers – are nervous just before going on stage. They rely on these nerves to give them the boost of energy which makes them sparkle. And the shot of adrenalin they get becomes a fix. It is something they can't do without, and is probably why these people love the stage experience so much. Nervousness is a useful, and essential part of performance, not something to worry about or be ashamed of. The art of effective speaking is not ceasing to be nervous; it is using the nervous energy to improve the talk. Standing up and speaking requires a great deal of effort: the slight lift given by nervousness arouses our energies.

If you feel you have an unusually nervous disposition, you may be surprised to know that you are not alone. Such sensitivity is common; psychologists calculate that: 'Between five and eight per cent of the population are unduly anxious.'[1] Knowing that you are not alone doesn't change the fact that you are nervous, but it should give you hope that your nervousness can be conquered. One of the more unpleasant features of being very nervous is a sense of isolation, and the fear of shame if others see that you are nervous. Take heart, there is nothing especially unusual in being highly sensitive, and you are far from alone. Almost certainly, there are compensating advantages in your higher than average levels of arousal, and sensitive response to anxiety. Highly nervous people, for instance, are often of above average intelligence. It is possible to apply this intelligence to solving the problem of nervousness by learning about it, and applying the results of research. The higher sensitivity is also compensated by greater alertness, and awareness of audience reactions. It sounds paradoxical, but is none the less true. Nervous people usually make good speakers, once they have tamed and applied their nervousness.

Research on nervousness

What is the evidence of research, and why do speakers feel nervous? Other people inevitably affect our psychological state, producing increased levels of arousal. The presence of others is natural, and solitary confinement is a terrible, disorientating, punishment. But too many people, and the feeling that we are being observed by others is disturbing. Performance quality, generally, is affected by the presence of others. Early in this century, Auguste Meyer found, in 1903, that when people were asked to do mental and written arithmetic, memorize nonsense syllables, or completed sentences in front of an audience, there was a 30-50% improvement in speed. Surprisingly, too, they made fewer mistakes.

More complex research was carried out by Dashiell, who used three different tests: multiplication, a mixed relations test, and serial association. Dashiell got his subjects to work alone, to work watched by others, simply to work together, and, last, to compete with each other. The result, in general, was that working with others present improves speed but diminishes accuracy. Other people make a difference to the performance of an individual. Quantity increases, but quality is reduced. The explanation is probably that speed of work, and the amount produced are obvious to the people watching, but the accuracy and quality of the work are less visible to onlookers. The individual in our society wants to excel, so he goes for what can be easily seen.[2]

Being observed by others affects our behaviour in well-understood ways. 'Arousal' is a technical term in psychology for a state of being 'keyed-up', marked by physical symptoms such as increased heart and breathing rates, dilated pupils, increased adrenalin, and alert and rapid reactions. Nobody is more exposed than the speaker, and it is not surprising that he or she is in a state of heightened arousal. While this is necessary to cope with the extra demands of speaking, if the arousal becomes too much the speaker starts making mistakes. The arousal is produced by the presence of the audience, and the anxiety by the fear of making mistakes. A competent, even merely satisfactory, performance reduces stress, since being observed by others is only over stressful if you are making mistakes. So your first aim must be to give the information clearly and adequately, then the anxiety about being observed will be replaced by a sense of achievement. If you are highly nervous it is essential not to try for too much too soon. Be satisfied with a minimum performance; there is less chance of mistakes.

Michael Argyle explains: 'The performer is aroused and anxious because his esteem and image are exposed to the risk of being damaged.' Of course, the level of arousal will reflect the risk, and the importance of success. Argyle continues: 'A well-known law in psychology states that increasing arousal has an energising effect, which first improves performance, but later leads to deterioration, as emotionality disrupts the pattern of behaviour.'[3] The art is to achieve an optimum level of arousal. Experienced speakers claim that their nerves are useful to them – it keys them up to give their best. As I have already said, people who make their living by performing in front of an audience often rely on the boost of adrenalin for the added energy. But the inexperienced speaker rarely has the problem of not being nervous enough. How does he cope with his nerves?

There are a number of factors which affect the level of nervousness. These are the size of the audience, the importance of the audience, how familiar the speaker is with the members of the audience, the difficulty of the subject, the experience of the speaker, and the vulnerability of his or her public persona. Firstly, the size of the audience is important. Twelve people is less daunting than two hundred. If you have to give a presentation at a large conference next month, it will help to give a trial run to a dozen colleagues this month.

All the factors which increase arousal add together, and size can be compensated for by practice with the same subject, as well as extra experience with speaking. Secondly, the importance of the audience increases anxiety. If important bosses, or people who can further your career, are present, try to reduce the overall anxiety levels by choosing a simpler subject, or chosing a more modest personality to project. You are unlikely to be promoted to the board as a result of your first talk; but if you make a hash of it, you may lose the chance in the future. Try not to aim too high, if the stakes are high. On the other hand, if you know all the audience personally, you can afford to set yourself a more difficult task.

The most anxiety-producing of all these factors is lack of experience. Even the first time crossing a high bridge can be nerve racking if you have vertigo, although if you do it regularly it will soon be a matter of indifference. So with speaking – excess nervousness is mainly caused by lack of experience. Get practice in undemanding situations, and on easier subjects, and future presentations will arouse less anxiety, though do not expect ever to be totally unmoved by the need to speak. The last of the factors which increase arousal is the risk of weaknesses in your persona being discovered. We all adopt a public persona, which differs from the one we show our family. It is not wise to adopt too demanding a persona when speaking: to pretend to be an expert on a topic when you are really a novice is inviting disaster. Equally, to try to be a smooth, casually competent speaker when you are as nervous as a bride at a wedding is foolish. It is the level of risk which pumps up the level of anxiety. Keeping the risk down will reduce the chances of exposure, and calm your nerves.

Fear of facing an audience is often simply fear of facing the truth about oneself. Exposing oneself to others invites them to see all those things which you secretly fear are wrong with your personality. In fact, few people are as bad as they fear, and most audiences are too interested in themselves, their own thoughts and their own concerns, to have time to do a detailed analysis of the speaker. Indeed if they did so they would probably expect the usual mixture of good and bad. They would only really dislike him if he were perfect.

Probably the best antidote to nervousness about one's own personality is the recognition that people who are unusual in their personalities or appearance often make a virtue of these features, and are liked and respected as characters. There is no need to be afraid of the identity you present to others. Only the fear itself is unattractive. All human variety is interesting and likeable; whatever your fears, if you are open and straightforward, the audience will like you for what you are.

The last section of this chapter presents a variety of antidotes to nervousness. They are not quack cures and many of them can be surprisingly effective. But before discussing them, let us look in more detail at the way nervousness affects a speaker and his or her audience.

Nervousness affects you

Nervousness has many effects on the body. The most obvious is the way that nerves affect the face. The muscles round the mouth tense up, and the eyes especially signal the wrong messages. Secondly, nerves affect the voice, especially affecting speed and pitch. Thirdly, nerves affect the performance generally. They cause timing errors: the biological perception of time depends on internal biological rhythms, which in turn depend on adrenalin levels. These rhythms go haywire when someone is keyed-up to speak.

Nerves also affect the way you think, interfering with language production, often resulting in ambiguity, and confusion of meaning. Nerves can cause mental blocks and loops. The only way to break a mental block, incidentally, is to abort what you are saying, and start again on a fresh part of the subject. When things are back on the smooth and level it may be possible to go back to complete the section which was abandoned. Finally, nerves affect the whole body, giving the speaker platformitis, jellied knee-caps, headaches, clammy hands, and an upset stomach. Nervousness is very unpleasant. It is one of the main reasons why inexperienced speakers dread the event.

One of the features of nervousness most obvious to the outside observer is the hesitation in the nervous speaker's voice. Many people think that 'ahs' and 'ums' are a sign of nervousness and some inexperienced speakers work themselves into a state of real anxiety trying to avoid ever saying 'um'. They fail, of course, and then let this failure embarrass them. Since it is such a trivial aspect of speaking, it is surprising to find how much research psychologists and linguists have done on this topic. Let me first explain the jargon: 'filled pauses' are 'er', 'ah', 'um', etc. 'Unfilled pauses' are silences. Lay and Burron chose a recording of a speaker with a high rate of filled pauses, unfilled pauses and repetition. These were edited out from a version of the recording.

Different groups of listeners scored the speaker on dimensions such as hesitancy, fluency, anxiety, tension and nervousness. These listeners judged the edited tape was more fluent than the original (though only slightly). But interestingly, neither female nor male listeners thought the speaker on one recording was more anxious, tense or nervous than the other.[4]

This is an interesting result, since hesitation is conventionally regarded as a symptom of anxiety, and speakers worry about these 'hesitation phenomena' as linguists call them. Yet it seems that listeners do not in fact perceive nervousness in the speaker when he or she hesitates. Hesitation seems to be less important than speakers fear. There are various kinds of hesitation phenomena: they can be divided into the 'Ah', 'Um' variants (also called 'filled pauses'); the 'repetition' variants, which include stuttering, repetition of sentences and phrases already spoken while thinking of what to say next, or while the mind is stalling; 'omission' which is where parts of phrases are left incomplete without realizing it; and 'sentence reconstruction' a variety of hesitation where after the sentence has been started the speaker goes back to restart it using a different structure. Finally, 'tongue slip' is where a non-meaningful noise or distorted syllable which cannot be recognized is emitted. Many studies have failed to find any reliable relationship between nervousness or anxiety and these phenomena. Mark Cook concludes that 'On balance, therefore, it seems that anxiety, permanent or transient, is not related to the use of Ah.'[5]

Speech rate has also been associated with anxiety, on the obvious theory that any anxiety will energize behaviour, and make the person talk faster. Research shows that what happens is not so much that the words themselves are produced faster, but that the pauses between the words are reduced. In a similar way, speaking slower consists of being relaxed enough to allow silences to grow between words and phrases. Interestingly, the results of research contradict the idea that nervous people speak more quickly when made more nervous. In fact 'instead of speaking still faster, they speak more slowly.'[6] In general though, the 'Ah' variant of speech disturbance was not correlated with anxiety.

The evidence is that listeners do not regard hesitations as signs of nervousness, although they prefer speech without too many pauses. Even good speakers have a scattering of hesitation phenomena; speech would sound unnatural, and rather plastic, without. The truth is that research on anxiety markers in speech has failed to show that filled pauses are interpreted by listeners as signs of anxiety.

Hesitation phenomena are not evidence of nervousness, or are not interpreted as such by listeners. But there are other grounds for avoiding too many of them; they impede directness, and the listeners' confidence in the speaker's grasp of his material suffers. But they should not be a source of worry; if they are, they are likely to increase, not decrease. They are entirely natural, and a moderate number of them probably signals natural relaxation, rather than anxiety, to the listener.

Nerves and the audience

The audience is disturbed by nervousness, as well as the speaker. There are two distinct ways in which the audience is affected; their judgement of the competence and subject knowledge of the speaker is affected by his or her nervousness (i.e. 'Why is he nervous if he knows what he's talking about?'): and their sympathy and concern are aroused by watching someone who is nervous ('The poor person is miserable!').

Firstly, the audience's judgement of the speaker's competence is affected by nervousness. The audience interpret the validity of the message depending on their perception of the assurance of the speaker. It is natural to feel that someone who knows what he or she is talking about, shows it in the confidence of his or her manner. So if a speaker is nervous the audience subconsciously feel it is because he or she doesn't know the subject properly.

There are two components to this: firstly, of course, it is difficult for an audience to realize how frightening they appear to the speaker. If a listener is sitting quietly in a chair, he doesn't feel very frightening! And he is not really aware of everyone else around him in the same way as the speaker is. So it is very difficult for him to understand why anyone should be nervous about talking to him. The listener tends to think that the speaker's nervousness must have some other explanation. Secondly, people who aren't telling the truth are often nervous. Whereas this isn't true of competent tricksters, and there are many other reasons why people are nervous, the unconscious effect of evident nervousness on the audience may be to make them suspicious. Consciously they may be sympathetic, underneath they find their confidence in the message undermined.

So nerves can affect the credibility of a speaker. Studies show that 'expressed confidence' (i.e. using confidence asserting phrases such as 'I am sure . . .', 'I have no doubt that . . .'), as well as confident behaviour, affects the amount an audience is persuaded by a speaker. It is also easier to listen to a speaker whom you believe to be an expert – there is a subtle sense of time well spent. Whereas listening to someone whom you suspect doesn't know what he or she is talking about is difficult, because it may be wasted.[7]

For these reasons, nervousness in a speaker affects the benefit the audience gets from a talk. The speaker's credibility is reduced if he is obviously nervous, and the audience enjoy the talk less. How do the audience know if the speaker is nervous? There are both obvious, and subconscious ways in which an audience perceives nervousness. The subconscious ways depend on non-verbal communication (see Chapter Nine); but also on a phenomenon which has only recently been discovered. Stress shows in a speaker's voice by signals which are beyond our conscious perception. Listeners are sensitive to the presence or absence of inherent micro-tremors in the speaker's vocal pitch. All voice patterns include an individual and unique level of micro-tremor (similar in many ways to fingerprints). When someone is placed under stress there is a marked drop in the frequency of vocal micro-tremors, which is registered by the listener. This phenomenon has been used to construct lie detectors, and it may explain why we sense if someone is telling the truth or not. To us it seems like a magic fifth sense, because we are unaware of the physical basis of the evidence, but through micro-tremors, we can judge just how nervous the speaker is.

As well as these unconscious channels of communication, there are many visible signs of nervousness. The basic sign is an inability to stand still when talking:

When a person is emotionally aroused he produces diffuse,

apparently pointless, bodily movements. A nervous lecturer may work as hard as a manual labourer. More specific emotions produce particular gestures – fist-clenching (aggression), face-touching (anxiety), scratching (self-blame), forehead-wiping (tiredness) etc . . . An anxious person tends to talk faster than normal and at a higher pitch.[8]

All these signs will communicate the speaker's nervousness to the audience. It is such signals which make a listener say, 'you can hear him sweating with thinking'. They can be controlled, of course, and they ought to be controlled if the audience is to be comfortable. Nothing is more distressing than seeing another person going through a purgatory of anxiety. Out of sheer kindness to your listeners, you should try to damp down the amount of random movement you make. Calmness in the speaker, even if created by conscious self-control, is reassuring and relaxing to the listeners.

Cures for nerves

Because nervousness is produced by purely psychological means, it can be controlled by purely psychological means. This is a point which many speakers have not realized. Bleeding when you cut yourself is a physical event, and requires a physical cure such as a bandage. Nervousness has real enough physical manifestations, such as sweating, feeling sick, and trembling. But it has a purely mental cause; bandages won't help nervousness, but ideas will.

In this section I am going to offer a series of ideas which will help you to see nervousness in perspective, and to control its effects. But in the end, the only cure for excessive nervousness is experience. And that is the most difficult thing to get if you are over nervous. The solution, as I suggested earlier in the chapter, is to set yourself less stressful speaking assignments for the first few times. As you gain experience, your nervousness will subside, and you will be able to face a large audience. But don't be ambitious first time out; learner speakers should drive carefully. And when making your first trial runs, remember the points made in this section. Each will reduce nervousness to a level where you are able to start to speak; increasing experience will then get the problem finally under control.

The first idea which offers a 'cure' for nervousness is the realization that the effects of nerves can rarely be seen from the outside. You feel dreadfully exposed when standing in front of an audience, but the plain fact is that they can't see what you feel inside; you are not made of perspex. It is almost always true that you look better than you feel. Like the ducks on the Bishop's pond, you may be paddling like hell underneath, but on the surface all appears calm. Remember that most of the audience are quite some distance away. Your eyelid may be trembling, your knee cap jumping like a jack-in-the-box, and your stomach churning like a steam engine, but none of this is visible from a few feet away. The back row can see nothing; even the front row can see little of what is really going on inside. So providing you prevent yourself pacing up and down, or waving your arms about randomly, you will appear to be calm, even if you are not.

Nervous speakers can rationalize their nervousness by thinking about the real situation they are in. Think about the audience as people, their motives, their hopes, and their interests; it will help focus your attention on realities, rather than your lurking fears. Here are six reflections which will help you gain this perspective:

1. It is an undoubted fact that an audience is made uncomfortable by a nervous speaker. There is a strong empathy between speaker and listener. One of the great showmen of speaking, Dale Carnegie, encapsulated this point in his dictum: 'I'm OK, you're OK.' Making yourself relax is a kindness to them as well. Think of yourself as helping them, and you will feel they are helping you.

2. Remember that the audience is *not* hostile. You were asked to speak, therefore they *do* want to know what you have to say. You are welcomed, since in effect, the audience has initiated the conversation by asking your opinion on a subject. They want to learn for their own benefit, and your job is to help. You also have the power of novelty, for they certainly haven't heard it before, at least not your way.

3. Remember that you are much more awake than they are, and much more self-critical. Therefore you are much more aware of errors and pauses than they are. What seemed like a dreadful mistake to you, was probably almost unnoticed by them. It may take them several minutes to become aware that something you said was peculiar. If you calmly correct the mistake, they will hardly realize you made it. Pauses, too, are perceived differently by speaker and listener. The audience is living on a different time scale, and what seems like eternity to the speaker may be barely noticeable to the listeners.

4. They are going to be more embarrassed than you, if the worst

happens and the talk collapses. It is only kindness to them, then, to keep going. Realizing that they are more frightened of failure than you are, makes it easier to be sensible. So try to keep the talk in order, for their sake.

5. An audience is naturally well disposed and sympathetic. Speakers are frightened of audiences because they imagine them to be composed of cruel ogres, who take malicious pleasure in failure, and sadistic delight in mocking errors. You may be surprised to know, if you are nervous, that this is not the case. Audiences feel involved with the success of the presentation, and the natural kindness of people is increased by their concern that everything should go well.

6. Even if everything does go wrong, they can't (and won't) actually shoot you. It's worth seeing your nervousness in perspective: what do you expect to happen if you make a mistake? The fact is that in many years of watching and teaching effective speaking I have *never once* heard derisive laughter. If the speaker is nervous, and makes mistakes, there is a sense of concern, and support from the listeners. The penalties for mistakes are very small, and most mistakes seem much bigger to the speaker than to the listeners, who may hardly notice. Don't worry: it isn't as bad as that!

In summary, one important cure for nervousness is to see what you are afraid of in a true perspective. Don't think of the audience as hostile and frightening: talk to them as individuals, and think of them as a collection of people. You would not feel that bad about talking to any one of them alone. Follow Machiavelli, 'divide and rule'. Remember that anxiety is usually at its peak just before you start talking. Once you are under way, you have to concentrate on what you are saying, and you forget about yourself. The keys are seeing the situation in perspective, careful preparation, and a realistic assessment of the audience. Providing you don't try to put on an elaborate front which you cannot sustain, nothing is likely to go wrong.

There remain, however, people whose misfortune is being over nervous, and who find simple rational self-control little help. In some cases this over sensitivity is genetic, in some cases it is due to bad experiences, such as too much hostility and teasing from school mates (perhaps because of a temporary problem – a stammer, a lisp, or a silly mother). Whatever the cause, there is no doubt that there are many people who cannot get on top of their nervousness by rationalization. They undoubtedly have an additional burden. Sartre once said that no one was born a coward, and everyone had the choice of whether he was going to be a coward or not. Nature endowed some people with a more lively sense of fear, and these people undoubtedly had more to triumph over in order to be brave. But nature had not made them cowards as such; that was solely, and only, their own choice. It is a stern lesson. If you are over nervous, it does not mean you cannot be a successful speaker, it merely means you have more work to do.

The very nervous

What can be done to help the over nervous? Clinical techniques have been evolved to help psychiatric patients, which may also be applied to normal people. Desensitization treatment uses controlled exposure in small, un-threatening amounts. The exposure is gradually increased until the potential for anxiety from an observing audience is close to a real-life situation. It is the way of acquiring experience gradually while being protected from traumas which would give that experience a negative effect on confidence. The easiest way to provide desensitization therapy is through a planned progression, from simple talks to more demanding ones. It helps to be frank, and tell colleagues and friends that you are trying to control your over nervousness by getting used to speaking. Try to arrange to give a short (10 minutes, no more) presentation to half a dozen people from the same office or laboratory. Then try a bigger audience, and gradually desensitize yourself.

The basis of the effect is this: anxiety produces a rush of adrenalin in a dramatic response to a threatening situation. If the situation produces nothing fearsome, then the adrenalin has been wasted. Next time the same situation is encountered, the body produces slightly less adrenalin, because it has learned that it is working for nothing. With increasing exposure to the frightening situation, the body learns to react less and less, as it learns that its response is mistaken. This process of learned response is quite involuntary, and cannot be changed once learnt. It is quite impossible to feel fear when driving a car if you are an experienced driver, though statistically it is the most dangerous thing you do in your life, certainly much more dangerous than speaking. Next time you are on a long, boring motorway journey, and find yourself feeling drowsy, try to feel nervous about driving. I'm sure you will not be able to. Desensitization has the same effect on nervousness about speaking. Eventually it will become as routine as driving to work.

If you cannot arrange a graded sequence of speaking assignments for yourself, or if you are very nervous and can't bring yourself to try, a course on speaking is often useful. Many organizations now offer courses in effective speaking: giving an exercise presentation to a group of other learners is much less threatening than the real thing, and being told when you do well, and helped to correct simple mistakes, will be just the boost needed to get you started as a competent speaker. Even reading this book, for reasons I will come to at the end of this chapter, is a form of desensitization, and will help you face an audience with more confidence.

De-stress techniques

There are other physical de-stress techniques, which some find helpful. For instance, high levels of adrenalin upset the body's chemistry by preparing it for violent action, which it doesn't get. Burning off energy by filling your blood with oxygen will help to replace this missing activity. A bout of steady deep breathing does this. Standing up and walking round also helps. The level of adrenalin can also be reduced by deliberate relaxation. Try the technique of clenching each group of muscles from your toes upwards. Tense each group in turn, hold for a count of five, and then relax them. By the time you get to the neck and mouth muscles (as well as the forehead group), you will feel considerably more relaxed.

One way in which people work off stress is to become angry; it is supposed to be the method used by some surgeons in the operating theatre. All these techniques will help deal with the overload of adrenalin, which is causing the problems. It is even possible to use chemical therapy to reduce the levels of arousal. Thus *Inderal*, among other anxiety reducing drugs, has been successfully used to provide the initial platform of confidence on which de-sensitization can build. Such techniques are only available under medical supervision, of course. None the less, if you are thrown into a major presentation, and are completely inexperienced and very nervous, a drug like *Inderal* prescribed by your doctor, may provide the help you need. Providing you do not become dependent on it for every presentation, there is no harm in a little assistance at the beginning of the learning curve.

The best cure for nerves remains experience. But how does a nervous beginner gain experience if he or she shies away from just those speaking situations which would gradually reduce this nervousness? It is possible to gain substitute experience through the imagination. Physical training instructors were surprised to find, in recent experiments, that simply making their trainees sit quietly with closed eyes and *imagine* themselves throwing and catching a ball actually improved their skills. What seems to happen is that imagining an exercise uses a similar process in the brain's balancing and perceiving circuits as the real act. The exercise helps these brain pathways to develop, and sharpen their responses. It is not only in literature that imagination can help real life. The same process occurs when preparing a talk.

By imagining the talk, the body responds as it would to the real situation. The prospective speaker feels nervous, and anxious. But each time he or she goes through the imaginary situation of standing up and opening his mouth in public, he has evoked another anxiety response which has proved to be unfounded. So each time the response occurs less strongly. Preparing a talk, going into the room which is to be used, and rehearsing the talk in front of a friend, reduces nervousness just as experience does. The process is more effective, the more precise and detailed the imagined experience is. Careful preparation is therefore very important.

Ill-prepared talks are often not only chaotic in content, but nervous disasters as well. The argument of this book is that if we understand the mechanism, we can control it. Careful and thorough preparation has this among other benefits; imagining just what speaking will be like reduces nervousness and increases confidence. This is why I suggested earlier that the simple act of reading this book would reduce nervousness. When reading about speaking, you are going through the processes of speaking in your imagination. At the end of the book, you will have increased your 'experience' of speaking, as well as your knowledge. This is not fanciful – research confirms that reading and thinking about something improves performance. Almost any preparation helps, but thinking carefully about task and audience is the best guarantee, for many different reasons, of successful presentation.

Notes to chapter six

1. Michael Argyle, *The Psychology of Interpersonal Behaviour* (4th edn., Penguin, 1983), p.215.
2. Dashiell, J.F., An Experimental Analysis of Some Group Effects,

Journal of Abnormal and Social Psychology, Vol.25 (1930), pp.290–9.
3. Michael Argyle, *The Psychology of Interpersonal Behaviour*, p.258 and p.21.
4. Lay, C.H., and Burron, B.F., Perception of the Personality of the Hesitant Speaker, *Perception and Motor Skills*, Vol.26 (1968), pp.951–6.
5. Mark Cook, Anxiety, Speech Disturbance, and Speech Rate, *British Journal of Social and Clinical Psychology*, Vol.8 (1969), p.14.
6. Mark Cook (1969), op. cit.
7. Catha Maslow, Kathryn Yoselson, and Harvey London, Persuasiveness of confidence Expressed via language and body language, *British Journal of Social and Clinical Psychology*, Vol.10 (1971), pp.234.
8. in John Corner, and Jeremy Hawthorn, (eds), *Communication Studies: An Introductory Reader* (Edward Arnold, 1980), p.54 and 57.

Further reading

There is also useful advice in:
Adler, R.B. *Confidence in Communication: Guide to Assertive Social Skills* (Holt Rinehart & Winston, 1977).
Daly, John A. and James C. McCroskey, *Avoiding Communication: Shyness, Reticence and Communication Apprehension* (Sage, Focus Edition, 1984).

Tension in performance

There has been a lot of research on educating performers to control their tension. You may find some of this interesting, and useful. If you want to read more about the control of nervousness, here are a few references:

1. Appel, S.S., Modifying solo performance anxiety in adult pianists, *Journal of Music Therapy*, Vol.13(1), (1976), pp.2–16.
2. Baird, F.J., Preparation, an antidote for stage fright, *School Musician*, Vol.32 (January, 1961), pp.34–5.
3. Barker, Sarah, *The Alexander Technique* (Bantam Books, 1978).
4. Ching, J., *Performer and audience, an Investigation into the Psychological causes of Anxiety and Nervousness in Playing Singing and Speaking Before an Audience* (OUP, 1947).
5. Grindea, C. (ed.), *Tensions in the Performance of Music* (Kahn Averill, 1978).

SUMMARY SHEET

Chapter six – Nervousness

Everyone is nervous, even experienced performers.

Nervousness is a valuable boost of energy.

Research shows that people work faster, but less accurately, when they are being watched.

Too much arousal disrupts a performance.

Inexperienced speakers should choose easy tasks first.

The size of the audience, its importance, the difficulty of the subject, and the personality you choose all increase nervousness.

Hesitations don't matter very much.

A visibly nervous speaker makes the audience nervous.

Don't show nervousness by pacing about.

Nervousness can be reduced by realizing that:
— it doesn't show from the outside
— audiences are sympathetic
— you were asked to speak, so they are interested
— the penalties for mistakes are very small.

Over-nervous people can be helped by desensitization.

Experience is the best de-sensitization treatment.

Reduce adrenalin by deep breathing and relaxation.

Tranquillizers can help.

Imagination is a form of experience.

Thinking and planning increases imagined experience, and calms nervousness.

7

Timing and bad timing

A contract

When people groan that they have been to a dreadful talk, the most common reason they give for their misery is, 'he went on and on, and on'. A poorly presented subject can be suffered, for the sake of the topic itself, if it keeps to time. But a talk which is both boring, *and* drones on for endless minutes after the clock shows that the finishing time is passed, is a torture. Even an interesting, well presented, talk which goes on for too long is remembered with little pleasure.

Timing is such an important subject, then, that it warrants a whole chapter to itself. No other aspect of the presentation can do as much damage to the way the audience remembers the talk; no other aspect is so easy to control, since it is a simple mechanical matter of looking at a clock face; and no other aspect is so easy to get wrong. Many people seem to think that timing doesn't matter; they have a casual attitude to finishing their talk on time. Talk after talk I have listened to has gone way over its appointed limits, and audience after audience has, not surprisingly, switched-off and ceased to listen. This chapter's sole purpose is to persuade you that timing matters more than any other single aspect of your presentation, and to show how simple it is to keep to time.

Why does timing matter so much? It is a question which I have thought about a great deal. It is quite obvious that speakers *don't* think it matters greatly. It is equally obvious, both from listening to others, and from observing one's own reactions when trapped in the audience for a talk which goes on far too long, that to the *audience* timing is vital. Why is there this difference? I have evolved three explanations for this, which can be summarized as *adrenalin*, *subject concern*, and *contractual*. Let me explain what I mean.

The first reason is the different adrenalin levels in speakers and listeners. Put quite simply, they perceive time differently. Among the various effects of adrenalin on the body, the ability to *endure* is one of the most important. In the wild, the ability to run for hours, or fight without tiring, is an important result of fear. In civilization, the heightened excitement and fear produced by speaking, causes adrenalin, the 'fight or flight' hormone to flush into the veins in large quantities. The result is that speakers have a stamina, a resistance to tiring, an endurance. which is superhuman. *They* can go on all day. It is this effect, too, which produces the strange pattern of elation and tiredness when you give a talk. Typically, you feel keyed-up and ready to go before the talk, and are totally unaware of growing tiredness during the talk. Half an hour afterwards, when the adrenalin level has dropped again, you suddenly feel worn out. While adrenalin artificially heightens the body's responses, you are unaware of tiredness. You draw on a physical overdraft of energy; afterwards this must be paid back.

The speaker, then, is in an abnormal state. He is indifferent to time and tiredness, and while he or she is speaking, they feel as if they could go on all day. But the audience is in quite the opposite state. Sitting down, and having nothing to do but listen, actually *reduces* adrenalin below its normal level. Listening is physically inactive; even the mental activity of talking to others is stopped. The audience, then, is at the other end of the scale from the speaker. The speaker has high adrenalin levels; the audience has low adrenalin levels. It goes a long way to explain why they have such different views on the passage of time.

Passionate absorption

The second reason is that audience and speaker probably have different emotional concerns about the subject. The speaker has been working on the topic for some time, preparing the talk. It may be his or her special interest, and it may be a lifelong fascination. It is quite common for the effort of preparing a talk about a subject to produce a quite passionate concern about the topic. I have seen novice speakers who had difficulty choosing a topic for a practice presentation, suddenly becoming great enthusiasts for the subject they finally settled on, button-holing people at coffee breaks and meal times to talk more about it. Speakers, like lovers, become emotionally involved in what they are doing. To give a talk on a subject is often to become a passionate advocate of that subject.

The audience, as usual, feels quite differently: their interest in the subject of the talk is unlikely to be so passionate. They may have no more than a polite interest in what is being discussed. They may have no interest at all in the subject, and have come to listen in the hope that the speaker will arouse an interest. Worst of all, they may have come because they had to, because they wanted to be seen there, because someone else (such as a boss) demands they should be present. They may have a passive, or negative, interest in the topic; even if they are keen, their interest is likely to be less than the speaker's. They will enjoy listening for a reasonable length of time, and then will want to do something else, have a break, or simply stretch and relax. They will certainly not have the overbearing passion for the subject speakers often feel while preparing and giving a talk.

The third reason for the different attitudes between speaker and audience is contractual. The timing of your talk is in effect a contract with the audience. You were invited to talk for a specific time, and you have agreed to talk for this time. The power of this contract is extraordinary. If you have been invited to give a ten minute presentation, the audience will become disastrously restless after thirty minutes. They will feel that the talk was disgraceful, that the speaker has committed some great social crime. But if you have been invited to give an hour's talk, and stop after thirty minutes, the audience will feel cheated. What you say may be no different on both occasions; the organization and effectiveness of what you say may not have changed, but the agreed length of time for the talk has changed. It is the contractual arrangements which are different.

There is no doubt that to over-run the agreed time is more disastrous than to under-run it. The explanation seems to be that the audience is quietly looking forward to the end of the talk. If that time comes, and passes, and the speaker is still industriously talking away, the listeners have lost their security. Now they don't know when they will be free to stretch, chatter, and do something different. If they were bored or restless, at least there was a fixed time when the torture would be over; when that time passes, and they still have to sit still and listen, there is no longer a time they can look forward to, typically they may think that another five minutes will see them free. Five minutes go by, and they are still imprisoned, and so it goes on. The level of frustration and irritation rises in intensity and becomes unbearable; a length of time which would be perfectly agreeable if arranged in advance, becomes a sheer misery when taken illegally.

By contrast, a talk which ends early is a welcome surprise. Listeners are keyed-up to listen for longer, and find themselves released pleasantly soon. If the talk ends when they think it should be half way through, they will be annoyed, but if it stops perhaps 10% short of its full time, they are likely to see only excellent organization, and praiseworthy brevity, in the speaker. Even five minutes over gives a bad impression, even five minutes under is splendid. Because of the way the listeners perceive time, you should always aim to stop slightly early; and if you find yourself going over the finishing time, the only tactic which will save the talk is to sum up in one sentence, and stop.

The explanation for this contractual imperative is simple. The speaker and the audience have opposed feelings and needs. The speaker is keyed-up, and emotional, the audience merely mildly interested. They are so different that only a treaty between them, which spells out exactly what each has to do, is viable. The audience have agreed to shut up, the speaker has an agreement to talk for so long. Audiences don't usually break their side of the bargain, but it is too easy for the speaker to break his or her side. If you talk for too long, the audience will certainly revoke their agreement to listen. Rather like a marriage, if you break your side of the agreement, the other side will break theirs. Running over time counts as irretrievable breakdown of the contract between speaker and audience. No amount of conciliatory behaviour will make up for it. Trying to be charming, interesting, even cajoling will not disguise the basic fact that you agreed to stop at 10.45 a.m., say, and are still talking at 11 a.m. You have a contract, which you must honour, if you expect the audience to listen.

A common excuse given for running over time is that the subject naturally needs more time. Speakers will say that they can't talk about 'X' in ten minutes, it needs at least 15 minutes, or that 'Y' needs an hour to explain, and cannot be presented in half an hour. This is nonsense; any subject can be briefly described in a few sentences, or discussed in detail over hours and weeks. It is necessary to be quite firm about this, because otherwise a speaker will always try to slip in a few extra points, or squeeze a few extra remarks under the net before the end of the talk. The rule is that the subject fits the time available, not the time the subject. It is the speaker's job to prepare his subject for the time he or she is given, and if a full description of the topic takes an hour, and you only have half an hour, then you cannot give a full description. A summary, even just a few keywords of the topics there isn't time to cover, is all you can offer. If you feel so strongly that your subject *must* have a fuller treatment, then your only recourse is to refuse the invitation to speak. There will always be someone else who can slim down the topic to fit the available time.

So for three reasons, the speaker's view of the task is different from the audience's. The speaker is wound up for endurance by his or her higher adrenalin levels, and fired by passion for the subject; but in the end it is a contractual conflict. The audience wants to know when the talk is going to end, and only have so much time to set aside; the speaker must honour this contract. One additional point: it helps to make your contract with the audience explicit, even if they already have it in their program. Start by saying, 'I have been asked to talk for ten minutes on . . .' It helps to remind them in this way of the listening task they are embarking on. It also focuses the speaker's mind on the timing, and it reassures the audience that the speaker *does* know what he is doing. It gives some hope that he or she will actually do what they have agreed to do.

The span of attention

The reason why talks are arranged for limited lengths of time is not only because people have other appointments, but because the average person's span of attention is limited. The simple fact is that about five to ten minutes is as long as most people can listen without a short day-dream. After a brief holiday to catch up with all the other thoughts floating round their head, people come back to the talk. But as time goes on these rests get longer and more frequent, and eventually it becomes impossible to listen any more. Research shows that this limit is around an hour, depending on the subject and the individual. That is why most talks and lectures are scheduled to last for an hour or less.

There are, then, physical reasons why talks are limited in length. The audience are simply unable to listen usefully beyond a certain limit. Research shows, too, that the span of memory is limited, as well as the span of attention. The plain fact is that the audience will not remember what you said if you drone on for long after the appointed finishing time, so no good purpose is served. Trying to force in more information, when the limit has been reached, is a waste of time. I suppose few speakers realize this obvious fact; once a talk has run over time no one is listening. People are looking at their watches, gazing angrily out of the window, shuffling their notes, tapping their fingers, even trying to tiptoe out: the one thing they are not doing is listening.

So the speaker continuing to talk long over his or her due time is an absurd figure. Nothing is listened to, and certainly nothing will be remembered.

It is worth elaborating this point, because speakers seem to think that if they talk louder, or faster, then at least some of what they say after time must sink in. No so. The mind has a remarkable talent for blocking things out. Modern psychology has confirmed the ancient philosopher's belief that the mind was not a passive vessel, to be filled like a jug, but more like a lighthouse, which only saw what it deliberately directed the beam of its attention towards. The fact is easy to illustrate; when you walk down the street thinking of something else, quantities of information are focused on the back of the retina, but only when something familiar is seen (like the number plate of your *own* car) are the firmly closed gates of consciousness opened, and the information is noticed. It is only what you *want* to see, or are interested in, that you perceive. Everything else passes by unnoticed.

How else could it be? We are surrounded every minute by objects, things to look at, information. Our minds would be filled to bursting in a few seconds if we saw and remembered everything. The mind was designed by evolution to perceive only what it needed, or wanted, to perceive. Our ability to ignore, reject, pass over information is probably our best developed mental faculty. It is invincible. Millions upon millions of pieces of information can be thrown at the human mind, and they will all roll off like water off a duck's back. We don't even get tired by the effort of rejecting information. It is natural and effortless. Every human being has a perfectly developed, unbreakable talent for ignoring things.

It is this impregnable fortress which the speaker is trying to breach when he goes on beyond his time. The listener has agreed to open the gates for so long. But listening is an effort. It requires actual mental energy to absorb information, even just to listen. It is as if the doors of the mind had to be held open against strong springs. After a certain time, the effort of holding them open becomes tiring, and the doors clang shut. Believe me, no amount of pleading, begging, or cajoling will make any difference then. Once people have ceased to listen, you are wasting your time; and they will cease to listen when time is out, and their agreement to listen expires. There is no point in going over time; and it is positively damaging to the audience's impression of your efficiency.

Inner time is bad time

No doubt many speakers know, consciously or intuitively, that little is to be gained by running over time. They probably feel, too, a responsibility, to their audience and their hosts, to stick to the timing they have been given. yet they still run over time. Why? The answer is usually that they have been using the wrong technique for time keeping. With the best will in the world, it is impossible to keep to time if you are using an unreliable clock. Many speakers use the most unreliable clock in the world, themselves. Human beings are marvellously adaptable mechanisms, and have many wonderful abilities, but one thing they are not good at is imitating Big Ben.

Of course, we do have many internal mechanisms which keep a rough sort of time. Daily biological rhythms help us to wake up at a regular time, and (hopefully) make us hungry only at meal times. We also have faster rhythms, like breathing and heart rate, which enable us to judge when we have been waiting for someone for 'hours' or minutes. Some people can judge half an hour quite closely when they are in a relaxed frame of mind. And that's the problem. When they are giving a talk, no one is in a relaxed frame of mind. Subjective perception of time depends on adrenalin levels, which effect all those body rhythms which we use to keep time. The heart thumps, breathing becomes tight, even the daily rhythms are upset when we have to speak. We may wake up early in the morning, and not feel very hungry at lunch just before speaking. Internal time is quite definitely unreliable when giving a talk.

The problem is multiplied by the fact that the unreliability is unpredictable. It is not possible to decide that the internal clock is running twice as fast, so one must speak for twice as long: or that the internal clock is running slow. It is simply haphazard. When one is nervous one enters a curiously timeless state, a suspended animation above and beyond time, where normal time has no meaning. This is why, of course, speakers often go on far too long – they don't realize how long it has seemed to the listeners.

What other techniques do people use to keep time? One common mistake is to imagine that a trial run with a watch will tell you how long the talk will last. You can't time yourself with the bathroom mirror, a tape-recorder, or a willing guinea-pig. An audience of one spouse, two children, and a dog will not produce the same adrenalin levels as the real audience; you will go faster or slower when you come to do the actual talk. As with subjective time, practice times are unpredictable. The reason is that in the heat of the moment most speakers think of something else to say, they elaborate a point, add an explanation, pause to unravel a difficulty. They also often find themselves missing out whole chunks of their material, when an earlier point diverts, and runs straight into a later point. The speed at which you talk is also different in real situations. Some people talk more loudly, and therefore more deliberately; others gabble nervously. The total effect is that a talk which took a perfect twenty minutes with the office tape recorder the evening before, may easily take 45 minutes in front of the audience. It may just as easily take twelve minutes! Believe me, I've seen it. There is no way to predict how much, or how fast you will talk in front of the real audience. The only mechanism which will keep to the same time in both situations is your wrist watch.

There is a specific technique for using a watch, but before I describe this, let me issue one or two more warnings about the timing of a talk. The first is simple: if you run out of time, talking faster in a desperate attempt to cram in the rest of your material is a waste of time. The audience will be shuffling, packing up their briefcases, and looking at their watches to see which train they can catch home. No-one is listening, so your hurried talking is wasting your time, and annoying them. Your talk will also end with an impression of irritation, rather than of triumph. Trying to make up for bad timing by a desperate dash at the last minute will not save the talk. It will ensure its total ruin.

If you do run out of time, you have to take brave and decisive action. Cut out everything that remains to be said. Summarize your conclusion in two sentences, and stop. There is no other technique which will work. One of the amusing popularizations of recent years was the application of the mathematical, suggestively christened, 'catastrophe theory' to every day situations. The mathematical idea was that, while apparently moving through one dimension, some situations are also drifting sideways in other dimensions. For example, what had been a simple slope in one dimension, if you try to retrace the same path, turns out to have a cliff edge. Trying to undo what has been done, or retreat from a difficult position results, not in an easing of problems, but in sudden catastrophe. The analogy is certainly true of international crises which have led to war, and seems to be just as true of speaking. When time has gone by, and you have reached the end of the allocation, it is too late to save the situation by going back. Time is irreversible. To try to say what should have been said earlier will only lead to catastrophe.

Absurd examples of catastrophic attempts to save a lost situation can be seen at some conferences. Here there are often large numbers of inexperienced and unwilling speakers, having their first airing in front of a large, and critically important, audience. In some conferences the microphone is switched off by the chairperson (after amber and red warning lights have been ignored by the speaker), in a last decisive move to stop a presentation which is running over its allowed time. Even this doesn't deter some speakers. I have seen resolutely erring speakers still standing at the rostrum, mouthing silently, after the microphone has gone off. The determination of some speakers to talk over time is certainly a strange phenomenon.

Good timing

Keeping to time, then, is vital. Using your subjective sense of time is useless; timing a dry run is just as useless. The only solution is to use a watch, or clock. But even this is not without pitfalls. In practice talks, despite the strongest warnings about timing, nearly a quarter of talks run over. When questioned, the three common excuses are these:

'I knew I only had ten minutes, but when I was half way through I realized I had forgotten when I started'.
'I added ten minutes to 11.35 a.m., and I thought I was due to finish at 11.50'.
'I worked out the finishing time, but then I got confused, and forgot it. I was trying to remember when I should stop all the way through the second half of the talk'.

Bizarre though it seems, all these are genuine excuses. Speakers, especially nervous beginners, find it very difficult to think straight when they are on their feet. I don't think this is a criticism: there is so much to think about when you are speaking, and your head is ringing with adrenalin and worry, that it is hardly surprising that you can't do mental arithmetic at the same time. To remember the starting time is all very well, but how do you add the length of your talk to it and get the right answer, without hesitating? The brain's processing power is swamped by the demands of keeping the flow of words going. There is no spare mental capacity to do even simple calculations once you have started to speak. The same is true of the memory. Your mind is straining to make sure that the right things are said in the right order. There is no space left to store the details of starting and finishing times.

I have even known speakers who admitted that they knew the time they should finish, but as it got closer and they still hadn't got to the end of their notes, they deliberately confused themselves, tried to believe they had made an error, and really had more time. The subterfuges of the mind under the stress of speaking are strange and wonderful. If these errors and tricks seem childish, and beneath you, you may be unpleasantly surprised when you do speak. What seems trivial and simple-minded advice in the cool quiet of private reading, becomes a life-line in the heat of battle. You may feel contempt for speakers who can't even add up while talking (which is like a certain President who was said to be unable to walk and chew gum at the same time). But unless a simple routine of recording time is followed, there is no guarantee that you yourself will not make one of these trivial, but disastrous, mistakes.

After many years, I have found that there is only one reliable plan:

Write down the *finishing* time in your *notes before* you begin.

Each word in this advice is essential. You will forget when you began if you rely on memory, so write it down. You can't do mental arithmetic while talking, so write down when you should *end*, not when you began. And write it down before you start, not half way through when you have forgotten. An alternative technique is to use a stop watch, or a digital watch with a stop watch facility. But you must still remember to start timing *before* you start talking. Whichever technique you use, adopt a simple routine which will become a habit. My tactic is to take my watch off and put it on the desk before I start. Just before starting, I check the time, add the length of the talk, and scribble the finishing time at the top of my notes. Then all the way through the talk I know where I am. If you adopt this simple scheme, you will have an added sense of security when talking – and you will never ruin a talk by running disastrously over time. It is a benefit well worth buying with such a small piece of self discipline.

Two final pieces of advice: firstly, the best timing technique in the world is useless if you can't modify your material on the fly. It is quite impossible to predict how long points will take, and the only way to achieve perfect timing is to be able to cut out, or add, material as you go along. The worst thing to do is to simply cut off the whole conclusion of the talk as time runs out. The structure of the presentation then becomes a headless monster, and people may be confused when you stop. Material from the middle should go, not material from the end. The only satisfactory way to organize this is to have your notes arranged as layers of optional material. One good technique is to have notes in different colours, red for essential material, blue for the meat of the talk, and green for extra points to pad out the subject if you need to fill extra time. Having your notes on cards also helps, since a whole wad of cards can be turned over if time is running out in the middle. Always have some spare material, and some optional material, so that you can modify the content of your talk as you go along. There is no other way to achieve good timing.

The second piece of advice is to be explicit about the control and administration of the time during the talk. It encourages the audience's confidence if you indicate briefly as you talk that you are aware of the passage of time, and are moulding your presentation to fit in with it. Mention that you are half way through, or that you have only five minutes to go, as the case may be. The audience dread a speaker who drones on, with no indication that he is aware of time. A sense of confidence in the speaker's reliability, and grasp of the situation, is given by explicitly marking the passage of time. But this timing must be done honestly. Don't say 'finally' several times in the last few minutes of the talk, to try to defraud the audience of extra time they are not willing to give you. It is rather like a debtor keeping creditors at bay by continual promises of payment. Simply marking the time when changing topics is all that is needed; and when you get to the last point, tell the audience clearly, but only once.

Good timing is one of the most impressive achievements in a speaker. It gives the audience a sense of security, and it surrounds the speaker with an aura of competence. By contrast, bad timing leaves the audience irritated and angry, and marks the speaker as incompetent. The effect of the simple matter of timing is dramatic, and certainly out of all proportion to the simplicity of getting it right. Why do so many speakers get it wrong? I really don't know; probably because they are bores.

SUMMARY SHEET

Chapter seven – Timing

Bad timing causes dissatisfaction.

Speakers have high adrenalin levels, audiences have low adrenalin levels.

Speakers are passionately interested in their subject, audiences less so.

There is a contract – you speak for so long, they listen for so long.

Talks which end on time, or a few minutes early, are always welcome.

Any subject can be made to fit any length of time.

Audiences can only listen for five to ten minutes without wandering. An hour is the limit of attention.

Inner time is disrupted by high adrenalin levels.

Test timing during preparation is unreliable.

If you run out of time, stop.

You can't do mental arithmetic while speaking.

You can't rely on your memory of times when speaking.

Write down the finishing time in your notes before you start.

Modify the content as you talk, to fit into the time left.

Tell the audience how much time is left.

Don't be a bore, and go on for too long.

8

Intonation and variety

Variety

Every speaker is passionately interested in what he or she is saying: and every listener is interested in many other things, as well. I suppose most speakers think that every word they utter is heard by everyone in the room. When you speak, it is hard to imagine that everyone else is not listening, too.

But the facts are different. The best evidence I can offer, is to think about your own experience as a listener. We know, from inside our heads, when we have been listening and when we haven't. From outside it is less easy to see, for nothing may change on a listener's face when his mind silently changes subject. But we all know that when we listen to a talk, we are not listening all the time. Perhaps you thought that you were unusual, inattentive, in some way bad, and never mentioned to anyone that your mind wandered off sometimes when listening to a presentation. I used to think that my lapses of attention, when I suddenly came to with a jerk and found I had been thinking of something else for a few minutes, while the talk had gone on without me, were some personal failing. But I have since discovered that everyone does the same.

An audience's average span of uninterrupted attention is perhaps only five to ten minutes. This statement may sound extreme; but the fact is that we all day-dream briefly every few minutes. Other factors, too, make attention a less simple and reliable component of the communication equation than we would like to think. One reason for the unevenness of attention is the big discrepancy between the number of words per minute a human being can speak, and the number of words per minute needed to fill our attention. The mind can process words much faster than the mouth can pronounce them;

rather in the same way that the Input/Output of a computer is much slower than the electronic processes inside it. There is therefore not always enough information content in spoken language to absorb the whole of the listener's attention. The effect is that the speaker must provide other stimuli if he wishes to hold the attention of the audience. If he does not their minds will wander: nature abhors a vacuum, and this is true even of the most vacuous minds.

One of the ways in which this hunger for enough stimulation manifests itself is the ease with which people let themselves be distracted. If there is anything in the room (or out of the window) to look at, the audience will watch it, just because they need something to fill the spare capacity of their minds. But the vagaries of a listener's attention are not something irreversible and inevitable. The clever and effective speaker makes sure that this spare capacity is filled with other versions of his message, not with irrelevant distractions. The purpose of the chapters on non-verbal communication and visual aids, for example, is to suggest ways in which the message of the talk can be supported by other kinds of information, so that the listeners remain fully awake and attentive.

The greatest enemy to attention is not the audience's day-dreams, for they can always be interrupted if something interesting happens; the main risk for any speaker is becoming monotonous. The moment boring sameness creeps in, the listener's mind will wander. The only way to prevent this is to provide enough variety to engage their interest. In Chapter 11 I will discuss ways in which visual aids can provide this variety. But it is not necessary to show pictures to keep an audience awake; the good speaker does it by providing variety in his own voice and actions.

Try listening critically to a good actor on T.V. Do not listen to what he or she is saying, but to the way it is said. The actor's voice is an instrument of great flexibility, continually varying, continually exploring the range of expression. I am not suggesting that every speaker should turn into a professional actor (though good acting is part of every good speech). Indeed, the requirement to act is probably one of the main reasons why the more modest and reserved of us find speaking a harrowing experience. We are frightened of the need to display our personality, to control and manipulate the audience through the force of that personality. But every speaker can learn something from watching the professionalism of good actors. Watch not only for what an actor does with his voice, but also for the way it affects you, the listener. Notice how your interest and feelings are drawn after the changes in the voice pattern; notice how easy it is to listen to the tones of a good speaking voice; notice how the words are shaped and given meaning by the rise and fall of the intonation. You may not be, and may not want to be, an actor: but there is a lot to learn from their skills. The most important of these lessons is the value of introducing variety into your voice.

There is an explanation for this. Variety increases stimulation, and therefore increases arousal and attention. Television networks use variety in news bulletins to prevent the listeners' attention wandering. You may not realize it, until it is pointed out to you, but in a news broadcast, the audience's attention is carefully manipulated. Almost every minute there is a change of picture or of voice, which is one of the reasons why most people find the news easy to concentrate on. Few speakers will achieve these standards of variety, but some change every few minutes is a minimum if audience attention is to be held.

There are a number of what linguists call 'paralinguistic', or 'extralinguistic' variables, which can be changed at will in addition to the content and actual words used. These are things such as pitch, tone, timbre, and patterns of stress. These components of a speaking voice are as important as the letters (vowels and consonants – or what linguists call 'phonemes') we use to form words. We use these extra-linguistic variants to give shape and emphasis to our meaning, to express attitudes, and to recognize emotions in others. Listeners are so sensitive to these extra signals in the speaking voice that Knower was able to show in an experiment in 1941 that: 'emotions in speech played backwards were still recognized at better than chance levels.'[1]

An effective speaker must not forget the power of inflection to communicate commitment, enthusiasm, disdain or uncertainty. A speaking voice in which these patterns have been reduced to a mono-tonous flatness communicates only boredom. There are a number of simple components in speech variety , a vocabulary of intonation. These components can be incorporated in everyone's speech. But before discussing the main ways of doing this, I must first stress the overriding importance of being heard. The most eloquent speaking instrument in the world is of little use to the listener who has to strain to work out what is being said.

Clear enunciation

Intelligibility when speaking is the same as neatness and legibility when writing. No manager or administrator would issue a document with blurred, smudged, or faint characters. Neither should you talk with faint, smudged, or blurred enunciation. Your primary responsibility in communicating is to ensure that the physical encoding of the message is easy for the audience to receive and decode. In less technical terms, it must be audible. If the receiver of the message is using most of his mental energy to recognize the words being used, he will have less time for thinking about the content of the message. As with many of the basic requirements of effective speaking, this is such a simple and obvious point that it is often overlooked. Yet I would guess that three-quarters of the talks which fail, do so for the simplest reasons. And one of the simplest reasons is that it is unintelligible.

It is difficult to judge how clearly you need to pronounce the words, because you are closer to your own voice than the listeners. What sounds pedantic and laboured to you, standing on top of your own voice, may well be just clear and comfortable at the back of the room. An added problem is that you know what you are saying, and therefore recognizing the words is trivial: but others do *not* know what you are saying, and need to recognize all the words. Spoken language usually slurs over some sounds, and misses others (called *elision*), to smooth and speed the flow of syllables. This makes it difficult to understand foreign languages when they are spoken, for instance. Similarly, the attempt to get a computer to recognize English speech reliably has defeated scientists, despite large American research projects. For the same reason, if an audience is going to understand what you are saying, from the back of the room, you need to pronounce each word clearly.

There is no need to sound like a school-marm, but speakers do need to pronounce words rather more carefully than we typically do in everyday speech. Linguists have described in detail the vocal features that ensure comfortable and correct decoding at a distance. What are those features? I don't think a speaker needs a lesson in linguistics, but I do think he or she needs to have thought about the requirements of intelligibility. They are principally a matter of keeping the head up, sounding the ends of words, not dropping the voice at the ends of phrases, and not swallowing the vowels. All the major phonological features must be made fully articulate and explicit. The speaker must emphasize those features of the message which in normal (face-to-face) speech are conventionally glossed over in the interests of speed. Of course, the size of the audience affects how deliberate your enunciation must be:

A very large auditorium will have a decided effect upon the rate of speech – that is upon the amount of daylight needed between the words to permit the sound to carry to the farthest parts of the auditorium without serious interference from reverberation and the overlapping of sound. . . . Anyone who has had military training will recognise the devices that are used to obtain greater carrying power for audible signals. The infantry command 'right oblique' becomes 'rhit-ho-blhik'[2]

I am not suggesting that you try to speak like a army officer, but so many speakers forget the need for the voice to be clear at the back of the room, that a speaker should, in general, speak rather more clearly than he or she thinks necessary. The bonus of clarity will be gratefully received by the audience. There are simple physical reasons why the voice doesn't always carry well. Different frequencies, and different types of sound, carry over distance with differing efficiency. Many features of the voice, such as the stopping noises made by the tongue, and the 'unvoiced fricatives' which are mainly breathed, rather than sounded, all become difficult to hear at a distance. An example is the 'p' in jump'. Sounds like 'f', 'j', 'p', and 'x' may also be more difficult to distinguish; the difference between 'click' and 'lick', which depends on a faint liquid clicking noise before the 'l' in the first, and not in the second, word, may be lost.

There are many features of spoken language which depend on such small differences in the sound, and these cannot easily be magnified. Try, for instance, shouting the 'c' in 'click'! It is quite easy to increase the volume of the '-ick' component; but impossible to increase the volume of the 'cl-' component much. The solution is to increase the silence that precedes the 'cl-', so that it becomes more distinct. It is this kind of manipulation which is the essence of intelligibility. In practice language users are able to increase the clarity of their voice at will. What is usually missing is the will, or the knowledge, of what and why the voice must be made clearer.

Remember that in their struggle to hear, listeners may also have to contend with echoes. Few of us realize, too, how much we depend on being able to see the speaker's mouth. Lip-reading can contribute a surprising amount to the clarity of speech, and this is difficult from any great distance. So a certain percentage of the normal sound pattern is lost when we listen to a speaker in a large auditorium. This means that an audience may find difficulty in understanding unless the speaker deliberately saves those features which are slurred in casual and intimate speech. In sound, as in most areas of speech, language has a measure of redundancy. It is not necessary to hear 100% of the speech sounds in order to comprehend. But if too much is lost, understanding may demand large amounts of brain-processing on the part of the listener. He gets tired, and soon begins to miss larger and larger chunks of the message. If the speaker is aware that features which make the message intelligible from close by are lost at a great distance, he or she can ensure that they compensate.

Varieties of intonation

I have already said that a good actor uses variety in the intonation of his voice to create interest and emotional excitement. Normal speakers, too, use intonation to convey meaning. Try saying 'You are going tomorrow' with the voice falling at the end of the phrase. Now try saying the same phrase with the voice rising at the end:

> You are going tomorrow.

Without changing the words themselves, the phrase has been changed into a question, simply by the intonation. Intonation is a powerful tool in the speaker's armoury. A linguist summarizes the importance of intonation patterns like this:

> The pattern of pauses, stress and pitch is really part of the verbal utterance itself. Pauses provide punctuation (instead of saying 'full stop' as when dictating); stress and pitch show whether a question is being asked and provide emphasis, thus showing which of several possible meanings is intended. . . . The same words may be said in quite different ways, conveying different emotional expressions, and even different meanings, as when 'yes' is used as a polite way of saying 'no'.[3]

This system of intonation patterns is not just an auxiliary to meaning. It is often the determiner of meaning: it is also the way in which we give life and interest to our voice. When we say a speaker has a dead-pan voice, we are pointing to the absence of intonation variety in the voice. If words are spoken in a flat, monotonous, way they soon become boring to listen to. If you listen critically to speakers, you will find that the ones who are interesting, and easy to listen to, have a rich variety of intonation; the ones who are boring and dull have a level, unchanging intonation. The variety, flexibility, and mobility of the voice is one of the main keys to interesting talking.

The inexperienced speaker, or the person whose chosen field is technical or professional, is not likely to compete with Richard Burton, just by being told to increase the variety of intonation in his or her voice. But all is not lost: there are simple ways in which you can consciously vary your speaking voice. I have found over the years that paying deliberate attention to these features can transform a dull voice into an interesting one. The key is variety.

I distinguish six rules of thumb to introduce variety into a speaking voice, six dimensions along which the voice can be manipulated. You should not, of course, use only one, and, for example, talk alternately faster and slower throughout the talk, while maintaining the same monotonous tone. Nor should you work through the types of change one by one, while ignoring the content of what you are saying. The aim is to introduce variety, so that a natural expressive flexibility charms the audience. This is done by being aware of all six possible dimensions of change, and trying to remember to use all of them in various combinations through the talk, as seems appropriate. I can assure you that the result will be surprising, for I have seen these simple tactics transform the most tedious of speakers into someone to whom it is at least *possible* to listen!

1. Pause

Silence is a more important factor in speech than most speakers realize. Silence is, for instance, the main ingredient in a comedian's timing. The audience savours the carefully judged length of the pauses before the punch line is delivered. Similarly, in informative speaking varied pauses counterpoint the meaning. Silence is a powerful way of communicating; leaving a gap gives time for the meaning of what has just been said to sink in, and it clears the way for the importance of what is to come. But a nervous speaker unfortunately finds it difficult to leave silence. Terrified of the echoing pauses, nervous of losing the audience if he or she stops for even a moment, they rush breathlessly on, filing every nook and cranny of time with sound.

The result of continuous speaking is that the audience's minds become clogged with information. They are thinking over one fact, when the next one comes in, and then the next. There is no time to absorb, and soon the audience's minds are drowning in information. An experienced speaker, though, knows the value of silence. I do not mean long embarrassed pauses, when the speaker is struggling tongue tied for something to say, or shuffling through his notes in desperation. It is the carefully judged pauses, the fractions of seconds of meaningful silence, which tell. Nothing increases the impression of confidence and control as much as the ability to *stop* talking for a calculated pause, before a change of subject, a vital point, or a surprising fact. Try to make use of the pause as a deliberate item in the articulation of your message.

Some speakers are terrified of stammering, or even uttering a single 'er' or 'um'. They seem to think that a talk should be delivered as a continuous, smooth, glossy and uninterrupted flow of sound. Not so. As is pointed out in chapter six, all speech has hesitation phenomena: pauses, fillers such as 'er', 'um', and 'like', false-starts, repetitions, and even stammering are all natural. There is evidence that they make speech *easier* to listen to, as long as they don't become intrusive. They can also be effectively used as pauses to articulate the flow of information.

One of my friends is distinguished by a slight, but persistent stammer. Despite treatment, he continues to be hung up on a word every now and again. A handsome man, possessed of great charm, he makes the best of it, and has a firm and confident manner which openly admits his stammer, and refuses to be embarrassed about it. He has became a successful salesman, and has had a succession of strikingly attractive partners in his life. His stammer seems to raise interest and expectation in his listeners, and his speech has a remarkable emphasis and variety. Even stammering can be used to effect: no hesitation, if you make use of it, is necessarily bad.

New speakers are understandably nervous of any silence, but a firm grip, and a deliberate pause between each point, will help steady their nerves. It increases the audience's perception of the speaker's control of the situation. Silence is a great asset in a talk; make use of it to add variety to your presentation.

2. Pace

The second dimension along which the voice can be varied is speed. The pace of the speaking voice, along the whole range from slow and deliberate emphasis to rapid enthusiasm, can be consciously varied. Highly charged points can be made word by word; amusing anecdotes can rush on to their punch-line. It is usually easy to see where slowing down would be appropriate; most speakers talk too fast. The best way to change the pace of the presentation is to mark places in your notes where there is need for special emphasis, or deliberate clarity. When you get to these points, force yourself to slow down. When you relax the restriction your voice will rapidly regain its normal speed. Of course, if you normally speak slowly, you will need to deliberately hurry up from time to time. Conscious use of varied pace adds to the attractiveness of the speaking voice; monotonous regularity of speed increases the risks of boredom. It is, incidently, very rare for listeners to complain that a talk is going too slowly. The nervous speaker usually talks too fast. Not only is it tedious if the pace never changes, it also leads to over rapid unloading of information. Varying the pace can reduce the strain on the audience, as well as introducing a refreshing variety.

3. Pitch

Nervousness contracts and tightens the muscles around the throat and voice box. The effect is that the average pitch of the voice rises. The nervous speaker produces a typically 'strangled' sound, which is unpleasant, and tends to evoke anxiety responses in the audience. If you are aware of this problem, then the pitch of your voice must be consciously lowered. Pleasant actor's voices are usually low-pitched and 'gravelly', and actress's voices 'husky'. The hearers interpret this as representing confidence and calmness, although it may be nothing more important than the chance construction of the voice box.

Because it relaxes and reassures the listeners, and because it increases the variety of the talk, pitch should be deliberately varied, and usually lowered. It is not as difficult to control as many speakers think. A little practice in speaking at higher and lower pitches, perhaps with a tape-recorder, can soon educate the voice into a more conscious flexibility. Try repeating the same word into the microphone at various pitches: for example say 'tomorrow' in low, medium low, middling, medium high and high pitches. When you have practised this a few times, and listened to the results, try the same pitches with a full sentence. Say 'I'm going out tomorrow' in low, medium, and high pitches. When you have listened to this, and educated your voice so you can consciously control the pitch, try a longer passage. You might try saying: 'I'm going out tomorrow, and I hope the weather is fine, don't you?' It is almost impossible to say this without the voice rising and falling to follow the natural intonation demanded by the meaning. You will probably say something like:

I'm going out tomorrow, and I hope the weather is fine, don't you?'

This sort of practice is just like doing scales on the piano. It can be tedious, and it needs a lot of discipline. It's purpose is to increase your control over your own voice, and to make you more conscious of what you are saying. Like piano scales, it greatly improves performance. If you think such practice is pretentious, don't try it. But you *will* find that varying the pitch of the voice is a useful technique. For example, a well known trick which can have a dramatic effect is the sudden *lowering* of the speaker's voice to emphasize an important point.

4. Tone

We all recognize an angry tone of voice, a worried one, or a confident one. It is too detailed for our present purpose to analyse the exact components of these tones. In any case most people can pretend to anger, or other emotions, when needed – when sending the children to bed for instance. For most people, an awareness of tone is not difficult to develop. It is one way of introducing variety into a voice. Try putting directions such as 'sound surprised', 'sound pleased', 'sound concerned' beside relevant points in your notes. The talk will come to life in a surprisingly effective way.

5. Volume

The fifth source of variety is a change in the loudness of your voice. You should, of course, in any case be adjusting the volume of your voice to the size of the room in which you are speaking. There is a natural mechanism which controls the volume of the voice; a speaker adjusts the level of his or her voice, depending on what he thinks is needed. Greeting someone walking along the opposite pavement produces a volume of sound many times greater than intimate conversation. The speaker does not need to think about this consciously; it is one of the ways in which the brain automatically adjusts behaviour to the needs of the outside world. The sensible tactic for the speaker is to use this natural ability. What is it, then, which causes the volume of our speech to be adjusted? The answer is, awareness of how far away the other person is. If we look at someone some distance away we automatically adjust the volume of our speech to the appropriate level. Thus if a speaker *looks* at the back of the room the volume of sound he or she makes will be appropriate. But if he avoids looking at his audience (or only glances at the front row), the voice will be inaudible to the majority of the room. The volume of the speaking voice is also affected by assertiveness. Self-confident people tend to have louder voices. Shy people, partly because they fail to look steadily at the back of the room, have quiet voices.

Another physiological factor which affects the control of the voice is our own perception of our voice. We are usually unaware of how important this perception is. We hear our own voice in two ways: partly through the air passages connecting ear and throat (it is these passages which relieve the popping in your ears when you swallow as a plane climbs to cruising height). The second way is through vibrations in the bones in the head. This method of transmission greatly alters the *quality* of the voice we hear, and is one reason why people have so poor an idea of what they really sound like. There is also a third source of the sound of your own voice – reflection off the walls of the room you are in – though this is insignificant in a large room.

Since we are so close to the source of our voice, we often judge it to be much louder than it really is. It is therefore doubly important to look some of the time at the furthest person in the room in order to stimulate the production of a firm, clear voice. We adjust our speech from our own perception of it: 'It is easy to demonstrate how important to a speaker is the sound of his own voice. If his speech is delayed a fifth of a second, amplified and fed back into his own ears, the voice-ear asynchrony can be devastating to the motor skills of articulate speech'.[4] A cruel experiment, but it does demonstrate how important hearing our own voice is.

Once adequate volume is assured, the next element can be explored; loudness can be varied to give variety to the talk. Emphatic statements can be spoken in a louder voice. Relaxed discussion can drop back to a quieter voice. This contributes again to the relief of monotony, the great enemy of effective speaking.

6. Intonation

It is difficult to be aware of the exact intonation of one's own voice. There are two reasons for this. Firstly, we hear our own voice, as has already been pointed out, through the balancing air passages between mouth and ear, through the bone structures of our head, and partly through reflection from surrounding objects. None of these paths gives a realistic impression of the sound. Secondly, we know what we are intending to express, and therefore are unable to make judgements about the exactness with which our voice communicates that intention. We hear what we think we said, not what others hear. It is probably this fact, more than anything, which contributes to the lifeless monotony of some speakers' voices. They themselves hear variety, emphasis and intonation, whereas others hear only sameness. For this reason, the only really effective way to educate the intonation in our voices is help from others in the assessment, and improvement, of the way we speak.

Technically, 'intonation' is used to describe the way in which the voice rises and falls with the type of sentence being uttered. The voice rises at the end of a question, and falls at the end of a statement. But these changes are often small, and intonation can be increased by conscious attention to variety. The best aid to this is a critical and attentive friend but a good alternative is to use a tape-recorder. Most of us have some idea of what our voices sound like, but few of us have ever listened critically. Indeed, some people refuse to believe the sound of their own voice. I remember my first introduction to this strange phenomena. We had a lad at school who had a pronounced Yorkshire accent, which was sorely obvious in a south-western school. He was, in the humane manner of little boys, persistently teased about this. I don't think anyone ever spoke to him without a mocking music-hall version of a north country accent.

Being a man of spirit, even at twelve, he fought back. But it wasn't until I saw his reaction to the first tape-recording of his voice he heard that I understood. He was horrified. "But", he said, "it makes me to sound as if I've got a Yorkshire accent!" He was convinced that he spoke Queen's English. He obviously thought we were all being grossly unfair in teasing him, but doubtless put it down to the evilness of human nature, not to the reality of his accent. I have seen similar responses on many people's faces since; that is why it is necessary to be so carefully objective in judging the sound of your own voice. I recommend selecting a couple of passages from your prepared talk, or even reading from the daily paper, into a tape-recorder.

Listen very critically to the results, noting expressive features, and picking out uncomfortable features. Then read the same passage again, with an more varied intonation. You should be aiming to incorporate all the features mentioned: pause, changes of pace, flexibility of pitch, a meaningful tone, and a controlled and varied use of volume. Above all, try to inspire life into the passage. Make the words those of a living, feeling, being, not a speech synthesizer. If you practice this you will be surprised at how quickly your voice will acquire the precious variety that keeps audiences listening.

It is true that some people have naturally interesting voices. But they are few; and they usually become actors, not research scientists. One of the most outstanding faults in the 2,000 or so short presentations I have seen and discussed while teaching speaking has been the lack of life in the voice. Partly through tradition, partly through diffidence, partly through the wrong image of the intellectual man as unemotional, most managers and scientists speak in a dull voice. Careful attention to intonation will soon raise your standard of speaking skill.

The shifting sands of accent

A final point which I should mention under the general heading of intonation and variety, is the sensitive question of regional accent. One of the encouraging changes in social life in the decades after the Second World War was the acceptance of regional accents in intellectual and professional life. In the sixties, it became fashionable to have a regional accent, and many people adopted the accents of their upbringing, who would previously have disguised their voices in a veneer of poshness.

Linguists had a large hand in this change of attitude. Various studies demonstrated that regional speech was not a debased form of British English, but the remnants of older versions of the language. The historical and political factors which led to the domination of the accents of Southern England were shown up, and the absurdity of deciding on politico-economic reasons that one way of pronouncing a word was 'better' than another was widely accepted.

None the less, there remains some unsureness about regional accents, and if a speaker feels that his or her accent may be incorrect, it will undoubtedly interfere with his confidence. On the opposite coast of the USA, a New England 'twang' may be obtrusive, just as a Southern 'drawl' might be in Boston. But the moral is simple: do not worry much about the effect of accent; only try to interfere with your natural way of speaking if it is a barrier to the acceptance of your message.

It is useful to be aware of the phenomenon which linguists call 'style-shifting'. It is the tendency unconsciously (or deliberately) to alter our accents up or down what we think of as the social scale in order to fit in with other people's valuation. It is a sort of chameleon camouflage;

and has little more significance than wearing a suit to visit the bank manager, and jeans in the bar. It is dressing up the voice, putting a collar and tie on our accent for formal occasions, and a sweat shirt on our voice for the familiar. The public-school boy, working on the building site during the holidays, will style-shift downwards. The lower-middle class mother, visiting the doctor, will probably style-shift upwards, sounding rather posher than normal. These remarks are not intended as social criticism; they are a fact about the way language is used. Undoubtedly, when facing an audience, your style will be different from your style when telling jokes round the bar. If you are aware of this, and accept it as normal linguistic behaviour, you are more likely to hit the right tone. Everyone style-shifts, and accent, whether genuine or acquired, should not be allowed to interfere with confidence.

A different problem which affects the comfort of the audience is the aesthetic quality of the voice. An audience may be disturbed by the presence of undue mannerisms of pronunciation. Some people swallow the last syllable of each word, some tail off in volume at the end of each phrase or sentence, some have a nasal delivery, and some whistle like a kettle. Wilcox suggests that: 'If the 's' sounds tend to whistle, this can be stopped by a conscious effort to put less stress on the 's' sounds (this sentence, for example, will show up the whistling difficulty if there is one . . . Practising reading aloud a sentence like 'Wasps build nests on fence posts' will help.'[5] These are all minor points, and you certainly should not let your confidence be shaken by over-sensitivity. But it is wise to be aware, and to have thought about the various idiosyncracies which we all have.

Listening to your tape-recorded voice will help you identify any obtrusive phonetic characteristics. Better still, get a friend to listen. Explain that you want to know about any obvious mannerisms in your pronunciation which might disturb your audience. Then he or she will not feel shy about being objective, rather than kind. Incidently, the person you ask to do this critical listening for you must be a friend. That is, they must be someone you can accept criticism from – not necessarily your partner. As I said at the beginning of this book, criticism from others, based on objective observation of what you do, and frank reporting of the faults as well as the successes, is one of the quickest and best ways to improve your ability as a speaker.

Notes to chapter eight

1. Knower, F.H., Analysis of Some Experimental Variables of Simulated Vocal Expression of the Emotions, *Journal of Social Psychology*, Vol.14 (1941), pp.369–72.
2. Sidney Wilcox, *Technical Communication* (International Textbook Company, 1962), p.230–1.
3. John Corner and Jeremy Hawthorn, (eds), *Communication Studies: an Introductory Reader* (Arnold, 1980), p.57.
4. George Miller, *The Psychology of Communication: Seven Essays* (Basic Books, 1975), p.78.
5. Sidney Wilcox, *Technical Communication*, p.231.

Further reading

More information can be found in:

Hargie, Owen, Christine Saunders and David Dickson, *Social Skills in Interpersonal Communication* (Croom Helm, 1983).
Warm, Joel S., (ed.), *Sustained Attention in Human Performance* (Wiley, 1984).

SUMMARY SHEET

Chapter eight – Intonation

No one listens all the time, they need variety.

A monotonous voice causes day-dreaming.

Actors have flexible, changing intonation.

Enunciation matters because listeners are farther away.

Add variety to your voice by:
— pausing
— changing pace
— altering pitch
— using different tones
— modifying the volume
— giving your voice more intonation.

Regional accents are all accepted.

Style-shifting is common.

Check your voice doesn't have irritating features.

Listen to a tape-recording of yourself.

Listen to comments and criticism from friends.

9

Non-verbal communication

Communicating without words

We have been dealing up to this point largely with words, their organization, enunciation, and timing. But there is another system of signs which we use to communicate – the so called 'non-verbal' signs. These are all the hints, indications, and suggestions we communicate not by what we say, but by what we do. Many of these signals are very subtle: a half smile, a slight cough, or a sudden looking away are enough to tell people a great deal about what we are thinking. Erving Goffman distinguishes between the meaning that we 'give' in words, and the meanings we 'give-off' in non-verbal signals.[1] When giving a verbal presentation, we communicate not only with words, but also with a whole range of gestures, movements and expressions. It is these non-verbal messages, in addition to the verbal ones, which distinguish speaking from writing or telephoning. The physical presence of our body and the signals which it gives off make us 'speak with our vocal organs, but we converse with our whole body.'[2] It is, as Robinson says, as if the speaker 'is operating something akin to a symphony orchestra.'[3]

A doubter might argue that some speakers stand completely still, rarely look up from their notes, and have an impassive expression throughout their talk. It seems reasonable to ask in what way such speakers communicate with their bodies? The answer to the common sense doubter turns out, on reflection, to be surprising. Not only is non-verbal communication real, it is inevitable. It is impossible not to communicate:

There is a property of behaviour that could hardly be more basic and is, therefore, often overlooked: behaviour has no opposite. In other words, there is no such thing as non-behaviour or, to put it even more simply: one cannot *not* behave. Now, if it is accepted that all behaviour in an interactional situation has message value, i.e. is communication, it follows that no matter how one may try, one cannot *not* communicate. Activity or inactivity, words or silence all have message value: they influence others and these others, in turn, cannot *not* respond to these communications.[4]

So the impassive, unmoving speaker is communicating just as much as the mobile speaker, but the message is different. He is communicating his personality, and his attitude to the audience in an unmistakable way; if you have to listen to an impassive speaker, you have a vivid idea of what sort of person he is, and what it would be like to spend an evening with him. This idea comes entirely from the non-verbal signals which his behaviour is unconsciously communicating. The fact is that almost anything we do can express meaning: 'Besides ordinary verbal utterances, expressive actions obviously include gestures, such as nodding the head, pulling faces and waving the arms, but they also include such behaviours as wearing a uniform, standing on a dais, and putting on a wedding ring.'[5]

Roland Barthes, the French philosopher, in a famous collection of essays interprets even dress fashion as a system of signs which operates largely by contrast with past customs. The meaning of each detail chosen by the designer communicates by blending the conventional and the new. One interesting dress symbol is the man's necktie. A more useless piece of material is difficult to imagine, with its ability to catch food, and dip in soup. Yet there is no other piece of dress which gives such strong signals. The effect of giving a formal talk without a tie unless it is hot summer), or with a tie loosely slung round the neck (unless you are in America) is out of all proportion to the size of the piece of material. Even small details of dress are a major part of the system of non-verbal signals we use, whether we are aware of them or not.

Every aspect of behaviour signals something. Ever since the importance of non-verbal signals became accepted, psychologists have been experimenting on their effect on the way people understand messages.[6] All of this research confirms the importance of the non-verbal element of communication. Michael Argyle measured the way the non-verbal signals affected the listeners attitudes:

Non-verbal cues had 4.3 times the effect of verbal cues on shifts in ratings. . . . It appears that we normally use two channels of communication, verbal and non-verbal, which function simultaneously; conscious attention is focused on the verbal, while the 'silent' non verbal handles interpersonal matters, including feedback on what is being said . . . One advantage of inter personal matters being dealt with non-verbally is that things can be kept vague and flexible – people need not reveal clearly what they think about one another.[7]

As a result of work such as this, Michael Argyle developed a theory about non-verbal signals. Essentially, he suggests that 'language evolved and is normally used for communicating information about events external to the speakers, while the non-verbal code is used, by humans and by non-human primates, to establish and maintain personal relationships.'[8] The conclusion we must make is that the non-verbal signals are real, but tend to express attitudes to each other, and personal attitudes to the subject rather than factual information. They are therefore especially important in speaking, just because speaking is a personal relationship in which the attitudes of the speaker to his subject, and to his audience, are a major factor in effective communication of his material.

Non-verbal signals are unconscious

If non-verbal channels of communication are so important, why are we not all aware of them? The main reason, I think, is that we are so heavily word orientated, that we tend to undervalue other ways of communicating. Because our culture, and our schools, emphasize verbal ability so heavily, we tend to overlook the expressive possibilities of the non-verbal. There were non-verbal signals before there was language. Animals, after all, manage to negotiate their social lives entirely by non verbal signals. They make friends, find mates, rear children, work out their political hierarchies, and work together in groups, by means of non-verbal signals. The same is probably true of human beings.

I think that the importance of the non-verbal components in messages is one of the great discoveries of the modern human sciences. A large number of research projects in the last two decades show the special role of non-verbal messages. To pick out just two sets of results, one piece of research confirms that listeners are perceptive, and reliable, at judging non-verbal communication. Davitz found that using 37 messages demonstrating ten different emotions, 58 judges agreed on the emotion being expressed. Their agreement was much better than chance levels, and the emotion they agreed on, was the one the speaker intended.[9] In a second piece of research, aimed at estimating the relative importance of verbal and non-verbal components in communication, Michael Argyle and his colleagues showed that people gave over four times as much weight to non-verbal as they did to verbal cues. In another experiment in 1970, he calculated that non-verbal cues were six times as influential as verbal cues.

Another interesting result was that there were two differing interpretations when the verbal and the non-verbal cues were in conflict. When the verbal clues communicated friendliness, but the non-verbal cues showed hostility, the speaker was judged 'insincere'; when the verbal were hostile, but the non-verbal friendly, he was seen as 'confused'.[10] Albert Mehrabian found that the listeners overall impressions about a message gave only 7% of the weight to the verbal component. 38% of the judgement depended on the vocal cues, and no less that 55% of the weight was given to facial cues.[11]

Non-verbal signals, then, can undermine the verbal message. We must respect them. I have said enough, I am sure, to convince you, if you needed convincing, than the non-verbal component of a message is influential and inescapable. All this psychology is not just academic interest. The results give valuable clues to speakers, who can use them to improve the effectiveness of their work. For example, the result that conflict between the verbal and non-verbal clues leads to judgements such as 'insincere' and 'confused' warns us that we must be careful to match what we say with what we do.

Luckily, non-verbal signals are natural, and we need no training to produce them. We are all good at making these non-verbal signals, even if they are largely unconscious. Goffman says: 'we all act better than we know how.'[12] Non-verbal messages are subconscious in all except practised actors. Every speaker should not necessarily become a practised actor, although actors usually make good speeches, largely because they manipulate the non-verbal signals with professional skill. But every speaker should at least be aware of the effect of these signals. The knowledge will help you avoid making gross mistakes.

The first piece of knowledge the speaker needs is that we cannot prevent ourselves making gestures which inform the audience about our state of mind while we speak. The whole personality is on show when you talk, not just your knowledge and your style. Your character will be perceived and judged by the audience from what you do and look like, as well as from what you say. They will judge whether you are sincere, observant, thorough, flexible, and intelligent. Your background, education, and experience will all be transparent to them. It is worth understanding the system of signals an audience reads to make these judgements. We can then hope to offer an audience the signals we want them to see.

What you appear to be saying

As soon as they see the speaker, the audience automatically start the process of assessment. They want information about him, and they want to be able to assess the information about him they already have. There are many sources of information: all sorts of features of behaviour and appearance are 'sign-vehicles'. Watchers glean clues from his conduct and appearance which allow them to judge him.

The main elements we use as signal carriers are clothes. Sociologists find that: 'social class is easily signalled and recognised in this way. Accents are equally effective, and professional people may impress their clients by their offices and cars. . . . Hair has no constant meaning at all, but is simply an important area for expressing opposition to prevailing norms.'[13] Another sociologist, Edmund Leach, who studied primitive societies found that they were just as sensitive as our sophisticated Western societies, to nuances of social meaning. Every element in every-day life communicated meaning:

All the various non-verbal dimensions of culture, such as styles in clothing, village layout, architecture, furniture, food, cooking, music, physical gestures, postural attitudes and so on are organised in patterned sets so as to incorporate coded information in a manner analogous to the sounds and words and sentences of a natural language. . . . It is just as meaningful to talk about the grammatical rules which govern the wearing of clothes as it is to talk about the grammatical rules which govern speech utterances.[14]

Non-verbal signals are thus the foundation of our assessment of others. Appearance is perhaps most important when we present ourselves to a group. Any audience automatically uses these non-verbal signals to check, corroborate or question its reception of the verbal messages:

Knowing that the individual is likely to present himself in a light that is favourable to him, the others may divide what they witness into two parts: a part that is relatively easy for the individual to manipulate at will, being chiefly his verbal assertions, and a part in regard to which he seems to have little concern or control, being chiefly derived from the expressions he gives off. The others may then use what are considered to be the ungovernable aspects of his expressive behaviour as a check upon the validity of what is conveyed by the governable aspects. . . . The arts of piercing an individual's effort at calculated unintentionality seem better developed than our capacity to manipulate our own behaviour, so that regardless of how many steps have occurred in the information game, the witness is likely to have the advantage over the actor.[15]

Therefore what we appear to be saying is as important as what we do say. We must be careful to control it if we wish to succeed as a speaker. All types of movement communicate, not just gestures with the hands. Intonations of the voice, the clenched fist, the wave of the hand, the shrugging shoulders, and the lifted eyebrow are some of the more obvious ones. We all interpret this system of gesture unconsciously; and as Michael Argyle has shown, in interpreting the speaker's attitude, it is much more significant than the actual words used.

The conclusion is that the non-verbal code is as complex as it is important; it is difficult to disguise it, and impossible to silence it. We are what we seem, and it is difficult to seem something we are not. The best general way to cope with non-verbal signals, then, is to be quite genuine and open. If you are a strong supporter of the project you are presenting, admit it, for it will show anyway. If you are a member of the same group as those you are speaking to, don't try to hide it, for they will know anyway. This advice to be honest with yourself and your audience may seem strangely obvious, but it is surprising how many people seem to think they can deceive their audience into believing they have attitudes which they don't have. The first conclusion is that directness and honesty are best.

Dress signals

The second conclusion from an understanding of non-verbal signals is that what you look like will be a major part of your total message. In the light of this knowledge, the importance of personal appearance rises above the clucking advice to 'look neat' offered so often to novice speakers. The speaker is not doing as mother says: he or she is manipulating a channel of communication.

The question goes deeper than whether the audience will think you look a tramp, or a smoothie. It goes deeper, too, than remembering that the wrong clothes are distractions, and that you need to check the neatness of things like tie, collar, and buttons before starting. Certainly, if you have something wrong with your dress the audience will spend its time wondering whether you know about it, when you will notice, and what your reaction will be. Meanwhile they will not have heard what you are saying! But the real importance of the speaker's appearance is that it underlines the credibility of the message. If he or she misjudges dress, mannerisms, or actions that message will be torpedoed. Luckily, appearance is relatively easy to get right in our culture. Standard suits, or skirt and blouse, for instance, will communicate acceptance of the group norm in most business situations.

It is, of course, as fatal to overdress as to underdress. Most people realize this, and it is surprising how some hidden sense of belonging makes people in the same office or laboratory dress in a common way. One big computer company is notable because *everyone* wears a suit and a tie at work. Its competitors have unconsciously chosen a different norm, and everyone wears an open neck shirt and no jacket. There are no written rules, and I don't think there is anything sinister or conformist about these simple conventions. It is normal to feel part of a group, and requires deliberate antagonism to persistently dress in a different way. These norms establish themselves slowly, but insistently, in a group of people who work together.

In a talk, there is less time for norms to establish themselves, and therefore it is more important for there to be a conscious choice. If the speaker is dressed in the same way as the audience, he will be perceived as part of the same group. It is easier to listen to a colleague, and his presentation is more likely to be heard and understood. Imagine someone wearing jeans and a lumber jacket, talking about technical insurance matters to a group all wearing suits. What he said would have to be very good indeed to rise above the 'noise' in the communication channel created by his non-verbal signals. Just as bad would be someone wearing a suit, talking in a research laboratory where everyone wears open neck shirts. Of course, if non-comformity is an important part of your message, then go ahead; but at least you should be aware of the significance of what you are doing. And in truth very few technical presentations have such overriding sociological and political aims that it is necessary to make a strong non-verbal statement of objection to the group norms, along with the informational content.

An eye for an eye

Dress is an obvious non-verbal signal. But when we are talking to someone, or listening to someone speak, we notice mainly what they do with their eyes. Probably the majority of our impressions of someone come from watching what they look at. When we describe someone as shifty, we do not mean that their feet or hands keep moving about, it is the way their eyes move that we have noticed. A sympathetic face is almost entirely due to the eyes, and every natural gesture must be supported, or created by the eyes. A smile which is not reinforced by friendly eyes, is ominous or insincere. Sadness is mainly shown by the eyes, as is tiredness. When dealing with others, it is the eyes more than any other part of the face or body which we watch. Not just the duration of eye contact, but the speed and direction of eye movement all communicate.

The eyes are such an important component of non-verbal signalling between people, that when someone refuses to show their eyes, hiding their expression in downward stares, we think of them as uncommunicative. Eye contact is vital in normal conversation. Not, of course, a steady unrelenting stare, but passing friendly eye contact seems to be an essential component of assured communication between people. Most of us are intuitively aware of this, and the normal habits of non-verbal signalling through eye expression, and direction of gaze are well established. My concern here is not with day to day social intercourse between individuals, but with how the rules of eye contact should be interpreted when speaking to an audience. We must again use the evidence from psychological research to discover not only what is normal, but also what others understand from eye contact. We can then use this knowledge to be more skillful manipulators of the code when speaking.

Important work on eye-contact and direction of gaze has been done by Michael Argyle and his colleagues at Oxford. He summarizes his research:

> Gaze is a non-verbal signal itself, but a rather special one, since it does two things at once – it is a signal for the person looked at, but it is a channel for the person doing the looking. . . . When two are talk-

ing they look at each other between 25 per cent and 35 per cent of the time. . . . In addition to the amount and timing of gaze, the eyes are expressive in other ways:

Pupil dilation (from 2–8mm.in diameter)
Blink rate (typically every 3–10 seconds)
Direction of breaking gaze, to the left or right
Opening of eyes, wide-open to lowered lids
Facial expression in area of eyes, described as 'looking daggers', making eyes', etc.

It is found that glances are synchronised with speech in a special way. Kendon (1967) found that long glances were made, starting just before the end of an utterance . . . while the other person started to look away at this point.[16]

Gaze communicates powerful signals about the state of mind of the sender. As Ian Vine explains:

Looking at another may indicate attention, liking or threat depending on the duration of the look and the accompanying signals, and so on, while patterns of looking during conversations are intimately linked to the regulation of information exchange.[17]

Mutual looking can also express the degree of tension in a relationship. The greater the tension, the less the amount of mutual looking. As Kleck and Nuessle conclude 'eye contact not only indicates how attracted a person is to another or how tense he is interacting with that person, but it also is taken by observers to be a cue which can be used as an index of attraction and tension.'[18]

If we wish to appear normal and friendly, we must ensure that our pattern of gaze fits into what the audience recognizes as normal. When people are trying to dominate they look more, but once the relationship is fully established the lower-status person looks at the other most, especially while listening. In this and other ways, gaze is used as a cue for personality. People who avoid your gaze are perceived as nervous, tense, evasive, and lacking in confidence. On the other hand, those who look a lot are perceived as friendly and self-confident. Psychologists found that depressive patients avoid gaze to about the same extent as schizophrenics, and when they are interviewed they usually look downwards. Argyle, Lefebvre and Cook found that people were liked if they looked more, up to the normal amount, but that too much gaze was liked less.[19]

Such research helps us to understand why we, as listeners, sense so much about a speaker from the way he looks at us.[20] Our task as speaker is to employ this knowledge, so that we give off the signals we intend to send, rather than accidentally conveying an impression of shifty unease, when we feel only diffidence. Because the audience-speaker interaction is unfamiliar, our habits of interaction are not as practised as they are in face-to-face conversation. It is therefore too easy to go wrong, and by the simple matter of where we look, give people the wrong impression. Luckily, there is nothing difficult or mysterious about direction of gaze, and unlike some other non-verbal signals, it is fully under our control.

Look when speaking

There is one further, simple and obvious fact about direction of gaze which is often forgotten. We use our gaze to signal who we are talking to. When lecturing about eye contact, I often make a simple experiment. Looking round the room, I choose a tie or blouse whose colour I like. Then, looking at someone not wearing a tie, I say 'I do like the colour of your tie'. The person I am looking at is confused, the person I mean has no idea I am talking to them. It is a trivial demonstration, but it underlines the basic point that we understand that we are being talked to by whether the speaker is looking at us or not. It is normal to direct remarks, when standing in a group of people, simply by looking at the person we are addressing. A general question or remark is signalled by scanning round the whole group. As with so much of the behavioural information in this book, you can enjoy yourself next time you are bored at a party or meeting, by confirming the point. Watch what people do: it is quite unmistakable. Gaze is used more often than a name to indicate who is the intended receiver of the message.

What has all this got to do with speaking? Simply this: many speakers either never look at their audience at all, or concentrate their gaze on a few people in the front row. The result is that, consciously or unconsciously, the people in the back don't feel they are being spoken to. Listening habits are based on receiving messages only when we are looked at. Subtly, but uncontrollably, we interpret messages from someone looking elsewhere as *not* intended for us, and don't listen. But when the speaker *does* look at us, it is a powerful subconscious message that we must listen.

The moral for the speaker is simple. You must look at the audience, and you must look in turn at everyone. This means that you must not find one person in the audience, either a friend, or some individual who looks unusually sympathetic or harmless, and rivet him or her with a steady stare. Nor does it mean that you must flash your eyes continually round the room like a roving lighthouse, pausing for no one. The ideal is a brief fixation of varying length, moving randomly and naturally around the audience. This is best achieved if you feel yourself to be talking to a collection of individuals. Consciously try not to forget anyone; consciously try to relate to everyone as a person. The result will be a natural distribution of attention, and everyone in the audience will listen to you.

Of course, in a big room, it is probably impossible to talk to everyone as individuals, though great speakers seem to be able to achieve this effect. The less experienced speaker must at least look in the direction of the back row as well as the front row. It is too easy to talk only to those you can see, who are sitting immediately in front of you, but this must be resisted. I find a good technique is to interest myself in what people are wearing in the audience, and I then find myself looking from time to time towards the back as well as the front. A firm gaze, not dwelling for too long on any one individual, nor omitting any one, will communicate frankness and competence.

In summary then, the control of gaze is an important element in a presentation. The general principle is that you must look at all the people in the room, not just at the front row, or at a sympathetic friend. Look especially from time to time at the individuals in the back row, and those at the sides who are more difficult to see because you must turn your head. Do *not* look out of the window, stare at the ceiling, or the board, except during brief pauses for thought.

Talking to any group, you can gauge the reception of your message by watching the listeners. Concentrate on their eyes and mouths, which are the most expressive parts of the face. Experienced speakers always look round the room, and they seem to be able to achieve a natural and easy rapport with their listeners. If you observe carefully, you will discover that a large part of this rapport is created by the fact that the speaker is looking at everyone. You may not achieve such skill immediately, but some eye-contact is vital, and no matter how shy or nervous you are, you must train yourself to reproduce at least the outward appearance of the good speaker, if not his inner charisma!

Legs and bodies

The third component of the non-verbal signalling system, after our clothes, and the way we direct our gaze, is the signals given off by the rest of our bodies. Many of the more glaring faults of a poor speaker are shown by the way he either jumps around like an automaton, or drapes his body tiredly and languidly over the desk. Every aspect of your posture and movement will communicate a mental state to the audience. So don't pace or lounge, and avoid leaning, swaying, or dancing, as if you were tipsy.

As usual, psychology supports intuition, and what we understand subconsciously has been spelled out by researchers. Argyle writes:

> Attitudes to others are indicated, in animals and men, by posture. A person who is trying to assert himself stands erect, with chest out, squaring the shoulders, and perhaps with hands on hips. A person in an established position of power or status, however, adopts a very relaxed posture, for example, leaning back in his seat, or putting his feet on the table. Positive attitudes to others are expressed by leaning towards them.[21]

The best posture for a speaker, though, is neither aggressive domination, nor flippant self-assurance. The message a speaker usually wants to communicate is one of relaxed competence. This is best achieved by a natural posture, which expresses awareness, as well as control. Because naturalness is the essential component, I am not going to prescribe a certain stance. Indeed, it is natural to change position from time to time to stretch and relax.

An upright posture, back straight, and feet slightly apart is the most neutral way of standing. The hands can then be used freely to gesture, handle notes, and deal with any visual aids. Leaning on a table behind with the feet crossed, and the hands resting on the table edge just behind you, is also a posture which conveys sensible control. It is natural to move when a new point is being made in the talk; this provides a useful way of signalling the end of one subject, and the introduction of a new topic. My habit is to lean on the table while discussing the topic generally, and stand upright, facing the audience, for the summing up. But there is no set rule for posture. It must be alert and natural, no more. Fig. 9.1 shows some typical postures.

Many speakers manage to be thoroughly distracting when they are speaking. A speaker who sprawls across the table, or even worse puts his feet up, so the audience see his face between two large, dirty soles, is very difficult to listen to. His or her attitude does not convey mental alertness, and the audience are likely to feel little confidence in the accuracy and thoroughness of what he or she says.

Fig. 9.1 Good postures for the speaker

Equally bad is the speaker who races to and fro across the stage, making listening like being at Wimbledon. I think the imitation of a pacing tiger is probably the most common fault of positioning. Talk after talk I have seen has been spoiled by a speaker pacing nervously to and fro like a prisoner in solitary confinement. The motion is obviously soothing for speakers; it provides comfort just like sucking a thumb. They work themselves into a zombie-like state by marching to and fro, and work out nervous tension by covering great distances. But the speaker who is constantly in motion is very distracting for the audience. To begin with, his journeys from side to side of the room are so obviously pointless, and they express only uncontrollable nervousness, never purposeful energy. Don't pace to and fro!

Some speakers solace themselves, even though staying in one place, by shifting their weight rhythmically from foot to foot. I have even seen speakers lifting each foot in turn in a hypnotic motion, which has no connection with what they are saying. The only message it conveyed to the audience was that the speaker was unaware of what he was doing, and not in control of his body. Some speakers flex each leg in turn, as if they are doing a jig, or imitating a little boy in desperation for a toilet. Dancing on the spot is as distracting as pacing to and fro. Stand still, between small and natural changes of posture.

Almost any movement which is not related to what you are saying is distracting to an audience.[22] Some speakers play with their position in a distracted way, while mumbling on. I well remember a speaker who, while running through some complex mathematical analysis, used to lean his back on the board, and inch his feet forwards bit by bit, so he was leaning back ever more heavily. He was playing with his balance, and whenever he had to turn to write on the board, it was a great effort to regain the upright. One day, he went too far. Realizing all was lost, he kept talking, while his feet slid away in front of him, and his head slid slowly down the board, disappearing from view under the table. Everyone remembered the incident, but no one remembered what he had been saying! Speakers do not need to be clowns, and a still position, helps both speaker and audience to concentrate on the subject of the talk.

Giving the audience a hand

If you ask speakers what most worried them when they gave their first talk, quite a number will say that the problem they were aware of only after standing up to speak, was where to put their *hands?* If you have never given a presentation, it has probably not occurred to you that this might be a problem. But if you are self-conscious and nervous, when many eyes are fixed on you, you become acutely aware of the large white objects dangling at your side. Few inexperienced speakers use their hands to support what they are saying effectively, but nearly all good speakers use a variety of gestures to reinforce their message.

Why is it that hands become so important when standing in front of an audience, when they cause no problem when talking in a social group? One of the reasons for the importance of gesture is that the listeners' eyes seize on anything which moves. New speakers can be painfully aware that the audience is watching their hands, every time they move. It is probably a survival from our primitive forebears when any movement in the jungle had to be noticed immediately in the surrounding stillness. Any sort of movement in a speech draws the eyes of the audience after it.

There is no escape from the need to do something with your hands. Every position you place them in communicates a rich symbolism to the audience. Nervous speakers do strange things with their hands; I have seen speakers hugging themselves passionately (presumably providing substitute maternal comfort). I have also seen speakers working one hand round their back, and grabbing the other arm in an anguished half-nelson.

Equally ludicrous are the speakers who adopt gestures of authority from the days of silent movies. One speaker puffed out his chest, and thrust one hand firmly inside the breast of his jacket. Until it was pointed out to him, he seemed unaware that he was doing a music hall imitation of Napoleon, although the audience were enjoying it hugely. Even simple gestures, like the delicate pursing of finger tips, tapping one row on the other like Sherlock Holmes, seem comic when done absent mindedly in front of an audience. Perhaps the most common of all gestures, and the most depressing, is the speaker who clasps his hands in front of him like a penitent school child, and then, when he has got into his talk, starts wringing his hands like a sinner in anguish. The gesture is powerfully emotive, and never fails to cast gloom over the audience.

Another common gesture, is the speaker, male or female, who imitates a hair-dresser, and spends the whole talk nervously stroking hair or neck. It is a well-known symbolic defence gesture. It is also absurd if done repetitively throughout a talk. Another ploy is to readjust the tie all the time, or fiddle with a bit of clothing. Speakers in the throes of nervousness often don't know what they are doing, and are surprisingly repetitive. Another gesture which I have met repeatedly when tutoring speakers is the simple scratch, which becomes a rhythmic irritation. Some speakers lean so heavily on the table with both hands that it looks as if they are trying to stop it taking off. They have also been known to push a chair around the room like a toy train, and indulge in many other bizarre activities.

The solution to all such embarrassing absurdities is not simply to plunge your hands into your pockets. Hands in pockets look too casual, and can convey an atmosphere of insouciance. Most men soon start jangling small change, or bunches of keys. Ladies may have no pockets, but sometimes find some make-up, which they then fiddle with for the rest of the talk. Because speaking increases tension, there is an irresistible drive to fiddle with something, to divert the nervousness. It is uncanny how a speaker's hands will seek out something to play with, almost as if they had a life of their own. Watch speakers, and you will see hands moving inexorably towards anything that can be moved. A pen is a favourite object, or a visual aid pointer. Failing these, a sheaf of notes, a bunch of keys, or a glasses case. The objects become a mantra which is caressed, tapped from end to end, balanced in strange ways, or pushed to and fro like a child's toy. Be especially watchful, when you speak, that your hands are not fiddling. If necessary, get a friend in the audience to watch, and tell you afterwards so you won't do it next time.

A favourite object to play with are spectacles. I saw a speaker who underlined every sentence by earnestly sweeping off his glasses, and poising them in the air. After a pause, they were energetically put back on, only to be swept off again at the end of the next sentence. I suppose he thought that the 'Winston Churchill' look added force to what he was saying. In fact, it merely made him ridiculous. Any repetitive action will draw attention to itself. Anything which is mindless will convey the impression that the speaker is not in control of what he is doing. Be specially careful, then, not to solve the problem of what to do with your hands, by playing and fiddling with small objects.

Some people solve the problem by standing with hands on hips throughout the talk, but even this gesture can look like a sailor soliciting. To stand with the hands clasped behind the back looks like a school master. Hands clasped in front is the school-boy stance, or the saying-your-prayers stance. There is no escape from these messages. As I said at the beginning of this chapter, there is no way of not communicating with non-verbal signals, since behaviour has no opposite. The solution, therefore, is to accept the fact that standing in front of an audience requires you to use hands and body as part of your message, and to use your stance to convey calm control, and your hands to support what you are saying by gestures.

The grammar of gesture

The Italians, and other southern cultures, seem to find expressive gesture easy. The Anglo-Saxon is usually more inhibited, although if you watch an animated conversation in a pub, you will often see large amounts of descriptive gesture. But when standing in front of an audience, gesture seems to freeze, and it suddenly seems difficult to think of anything to do with the hands. There have been several attempts to invent a grammar of gesture, usually too complicated and more appropriate to a drama course than a speaking course. I have found, too, that too much discussion of the details of gesture tends to inhibit the spontaneous and natural expression of ideas by the hands. But some discussion is worth while since it will give you a few ideas to work from if you find it difficult to make your hands do any work.

The first advice is to avoid using limited and repetitive gestures. If you are going to use your hands (and there is little alternative) then they must be used clearly and deliberately. Many speakers seem afraid to make bold gestures, so make silly little movements. The most common error is vaguely waving the hands around at hip level. All this communicates is bashfulness. The simple rule of thumb is that unless your hands are still, they must be clearly visible. Like raising your voice, you should also raise your hands. My rule is that the line from elbow to wrist must be above the horizontal; the hands must always be higher than the elbows. This alone makes the gestures seem confident and descriptive, rather than timid and furtive.

Having the hands in view naturally encourages you to use them to make supportive gestures. But if you are one of the many people who do not use gestures normally, and who feel inhibited in front of others, the solution is to start simply; the easiest gesture is a simple description of the size and shape of physical objects. Many points have obvious accompanying gestures. Opposing points of view, balanced objections, 'on the one hand, and on the other' are natural and easy gestures with which to start. Once into the rhythm of gesturing, most people find it becomes second nature, and requires no thought. Figure 9.2 suggests some simple shapes gestures can make.

The psychologists have observed that the body is in continual use when speaking:

"On one hand . . ."

". . . and on the other hand . . ."

"So large . . ."

". . . on balance"

Fig. 9.2 Simple gestures

When a person speaks he moves his body and head continuously; these movements are closely coordinated with speech, and form part of the total communication. He may (1) display the structure of the utterance by enumerating elements or showing how they are grouped, (2) point to people or objects, (3) provide emphasis, and (4) give illustrations of shapes, sizes or movements, particularly when these are difficult to describe in words.[23]

We use hands to reinforce and aid the message. Shapes, for instance, can be conveyed much better with hand movements. There is plenty of evidence that audiences find it easier to listen to a speaker if he uses gesture; they understand the structure of points more easily from gestures. There is every reason, then, to use your hands to outline the shapes of things, show the balance of opposing arguments, and clarify the structure of an argument. It will also help you to feel less constrained and nervous, and make you appear confident and experienced. There are so many advantages to simple gestures, that they should be a normal component of every presentation.

Act at the right speed

Speaking, I have already suggested, is a form of acting, and the same conscious awareness of your whole body is needed. Movement of all kinds communicates, and all movement forms part of the pattern of meaning, even though the audience may not be conscious of its effect. Movements and glances communicate in a variety of ways; the speed, frequency and rate of movement of eyes, hands, arms, body and legs are all part of the message. A slow opening of the arms has a different meaning from a rapid flinging open of the arms. A slow stare has a different meaning from a passing glance. The speed and energy of the movement will communicate as much as its direction. A lazy, nerveless shuffle as you move about the rostrum with tell the audience as much about the quality of your personality, as the words you use.

The level of energy expended in movement is an important part of the message. Too little energy is sleepy, dull, and boring. Too much energy in nervous twitches, and random jerky movement, is worrying and off-putting. Try to keep to a natural level of energy output, appropriate to engaged, animated conversation. Move with control and assurance. Appear to be confident in what you are doing, and fully aware of yourself and your effect on others. Then your presentation will have greater impact, and the credibility of your message will be enhanced.

All the advice in this chapter about non-verbal communication may seem frighteningly complex. You may now have the impression that every twitch will tell against you, that the audience can read your inner thoughts with X-ray eyes. Not so. Indeed, it is surprising how little others realize about our true state of mind if we act in a controlled way. My intention in discussing non-verbal communication at such length is certainly not to make the speaker so self conscious and nervous that he or she becomes like a rabbit caught in headlights. It is merely to make speakers aware of the vast range of ways in which their information and feelings are communicated to the audience. There is no cause for alarm. As I said early in this book, most audiences are humane, and sympathetic to those who make a genuine effort. The risk is being unaware of the significance of what you do. Many speakers concentrate so much on the words, that they forget their bodies, and allow themselves to fall into absurdities. Knowledge, as so often, is 90% of the battle. If you know that the way you stand, move and use your hands will affect the way your message is perceived, then you will give it some thought. Thoughtful movement communicates competence and control, and the non-verbal signals you give off will be appropriate.

Notes to chapter nine

1. Erving Goffman, *The Presentation of Self in Everyday Life* (Penguin, 1971), p.14.
2. M.L.J. Abercrombie, *The Anatomy of Judgement* (Penguin, 1979).
3. W.P. Robinson, *Language and Social Behaviour* (Penguin, 1974), p.138.
4. John Corner and Jeremy Hawthorn, *Communication Studies: An Introductory Reader* (Edward Arnold, 1980), p.23.
5. Edmund Leach in John Corner and Jeremy Hawthorn's *Communication Studies; An Introductory Reader* (Edward Arnold) p.15.
6. See Druckman, Daniel, Richard M. Rozelle, and James C. Baxter, *Non-verbal Communication: Survey, Theory and Research* (Serge Library of Social Research, 1982).
7. Argyle, M., Salter, V., Nicholson, H., Williams, M., and Burgess, P., The Communication of Superior and Inferior Attitudes by Verbal and Non-verbal Signals, *British Journal of Social and Clinical Psychology*, Vol.9 (1970), pp.222–31.
8. Argyle, *et. al.* (1970), op. cit.
9. Davitz, J.L., *The Communication of Emotional Meaning* (Greenwood Press, 1976).

10. Argyle, *et. al.* (1970), op. cit.
11. Mehrabian, Albert, *Silent Messages* (Wadsworth, 1971).
12. Erving Goffman, *The Presentation of Self in Everyday Life*, (Penguin, 1971), p.80.
13. Michael Argyle, *The Psychology of Interpersonal Behaviour* (4th edn., Penguin, 1983), pp.200–1.
14. Edmund Leach in John Corner and Jeremy Hawthorn's *Communication Studies; An Introductory Reader* (Edward Arnold), p.16.
15. Erving Goffman, *The Presentation of Self in Everyday Life*, (Penguin) pp.18–20.
16. Michael Argyle, *The Psychology of Interpersonal Behaviour*, (4th edn., Penguin) pp.78–86.
17. Ian Vine, Judgement of direction of Gaze: an interpretation of discrepant results, *British Journal of Social and Clinical Psychology*, Vol.10 (1971), p.320.
18. Robert Kleck and William Nuessle, Congruence Between the Indicative and Communicative Functions of Eye Contact in Interpersonal Relations, *British Journal of Social and Clinical Psychology*, Vol.7 (1968), p.241, 243.
19. Argyle, M., Lefebvre, L., and Cook, M., The Meaning of Five Patterns of Gaze, *Eeuropean Journal of Social Psychology*, Vol 4 (1974), pp.125–36.
20. See also Beattie, Geoffry, *Talk: An Analysis of Speech and Non-Verbal Behaviour in Conversation* (Open University Press, 1983).
21. Michael Argyle, *The Psychology of Interpersonal Behaviour*, p.42.
22. See, for examples, Bull, Peter, *Body Movement and Interpersonal Communication* (Wiley, 1983).
23. Michael Argyle, *The Psychology of Interpersonal Behaviour*, (4th edn., Penguin), p.41.

Further reading

Two entertaining recent accounts of non-verbal communication are:

Goffman, Erving, *Interaction Ritual: Essays on Face-to-face Behaviour* (Pantheon Books, 1982).
Pease, Allan, *Body Language: How to Read Others' Thoughts by their Gestures* (Camel Publishing Co., 1984).

SUMMARY SHEET

Chapter nine – Non-verbal communication

Behaviour has no opposite.

Non-Verbal Communication (NVC) signals feelings and attitudes.

Audiences are sensitive, and accurate receivers of NVC.

NVC tells listeners whether the speaker is sincere.

A speaker's whole appearance and actions are signals.

Dress signals support, or opposition, to group habits.

Gaze is the main signal.

Speakers who don't look at the listeners are 'shifty'.

The person you look at, understands you are talking to him or her.

Don't pace, lounge, or dance; stand still.

What will you do with your hands?

Natural gestures, with hands higher than elbows, are best.

People understand better if you use your hands.

Don't move too fast – or too slowly.

Avoid absurd, repetitive fiddling.

Controlled movement communicates competence.

10

Arranging the physical environment for a talk

Physical comfort

Human beings are physical organisms, and whether we like it or not, we are affected by the physical surroundings we are in. The profession of architecture is based on this fact, but physical surroundings affect us in more ways than aesthetic pleasure. At the extremes, actual discomfort clearly makes any sort of mental activity virtually impossible. True, we can continue to think in remarkably unpleasant circumstances, but only for short periods, and only when we are highly motivated. For most purposes, intellectual curiosity, general attention to matters which are not immediately important, and the normal rules of courteous attention, disappear very quickly as discomfort rises.

The range of temperatures within which we are neither too cold nor too warm is very narrow. There are absolute limits beyond which life does not exist. The limits within which intellectual life is possible cover a much more limited temperature range. Any kind of discomfort usurps our attention, and thinking or listening to others ceases almost immediately.

Worse still is the subtle impact of the conflict between the physical and the mental. Human beings evolved as active creatures, able to hunt, forage, and escape. Consequently, blood vessels and nerves cease to function properly if we rest immobile for too long on a hard surface. The only time most animals are still is during sleep; and forced inactivity often results in sleep.

To all these discomforts, human beings add their own special brand of fragility. We are acutely conscious of others near us, especially if they are behind us. Civilization has many rules for protecting and reassuring ourselves, but the lack of ease still lurks. Our interactions with others hardly ever cease to require wakefulness, and produce exhaustion if too stressful, or too continuous. Many of us even add to these burdens by an acquired dependance on stimulants such as caffeine in coffee or tea, or even the narcotics in tobacco and alcohol. Undoubtedly, human beings were not designed to sit still and listen to others talking.

This book has not suddenly turned into a basic biology textbook. I only rehearse these obvious conditions of all our lives because others all too often forget them. Because a speaker is fully aroused, keyed-up with adrenalin to cope with stress, he is fitted to be unaware of physical surroundings; but his audience is not. It is they who suffer if the surroundings become uncomfortable. Since the talk is only being given at all for the audience's benefit, unawareness of their simple physical needs is, to say the least, counter productive.

Few speakers allow their audiences to undergo actual pain, although there are undoubtedly some. But many speakers seem unaware that within the range which avoids pain, there are still wide variations. Minor changes of temperature, and physical and psychological comfort can have large effects on mental alertness. These effects are often unconscious; they are also nearly always wrongly understood. Thus a talk given in a fresh room with a pleasantly cool temperature, good lighting, and comfortable, sensibly arranged seating, might be judged to be excellent. While exactly the same presentation in an uncomfortable, or stuffy room might be judged to be tedious and uninteresting. The effect, within these finer limits is rarely attributed to the right cause. All the more reason for the speaker to get the physical conditions right, so his talk gets the best chance of a good reception.

Luckily, physical conditions are usually easy to change. All that is required is awareness, and a few simple arrangements. Why does every speaker not make these arrangements? As I have sat through miserably uncomfortable talks, I have often wondered about this. The conclusions I have come to are simple. First, that speakers simply lack the knowledge that it makes any difference. They may feel the hot day, or stuffy air, but think that it doesn't matter. Secondly, the British attitude seems to delight in stern perseverance, whatever the conditions. The stiff upper-lip cannot take note of trivial things like hard chairs, and dim lights. I really think that nothing else can explain the apparent belief that there is something disgracefully namby pamby in caring for comfort, whether one's own or the audience's. Thirdly, speakers are often inhabitants of a different world while they speak. Wild-eyed with adrenalin, seeing the audience as through the wrong end of a telescope, far removed from every care by their passion for their subject, they feel themselves mentally above ordinary cares. The stiffling heat, the roaring road-breaker outside, the flickering fluorescent light, are all unnoticed as they plunge forward with their presentation. But the audience notices. The ability of the listeners to hear and understand gradually leaks away, as discomfort rises.

Speakers often ignore the physical discomforts of their audiences, yet this is one of the most important final preparations for an effective talk. After everything else is done, the simple physics of the room can undermine the chances of success. Every detail of the arrangements is subconsciously interpreted by the audience as a deliberate message. Since the speaker is the main originator of messages in the room, he is blamed (or approved of) for every discomfort or inconvenience. In a sense, this is fair, because the speaker can usually control most of the arrangements in the room. Few organizers will rebuff a speaker's request for a particular arrangement of chairs, ventilation or lighting.

Proxemics

The first element in the physical arrangement of the room is how close the speaker should stand to the audience. In arranging the chairs, and the position he or she will stand, the speaker must ask questions such as, Can they see me? Can they see my face fully? Can they also see my hands? And what about my feet? If a speaker is fully visible it is easier for his body language to be interpreted. As with enunciation in verbal language, visibility is the basis of non-verbal language. If the audience cannot see the speaker clearly, they will not hear so clearly either. Research shows that it is much easier to understand the speech of someone you can see, than of someone who is invisible.

The question of how close to the listeners the speaker stands is part of a study called 'proxemics'; a branch of linguistic science which describes the effects of nearness or distance in signalling attitudes. Thus, in a small seminar of up to a dozen, you can sit the same distance from the audience as they are from each other. The effect is to make them perceive the speaker as part of the group, and to stimulate relaxed discussion. If the speaker marks his separateness by sitting at a distance from the group, the effect is to make him seem remote, formal, and unfriendly. It will freeze the atmosphere. Another useful technique in a large lecture theatre, is to walk among the audience, which will give a sense of belonging to them. The art is to use the non-verbal signals which nearness gives, to reinforce the message you want to give.

Proxemics is a fascinating subject. One might think that there was little to say about a subject as simple as how far people stand from each other. The truth is different: much interesting work has been done. Quite unconsciously, people adopt positions which signal their attitudes. You can amuse yourself watching people's social stances on the next occasion you have time to kill waiting in a public place. People adopt positions standing at 90 degrees to each other, and at a uniform distance, depending on the degree of intimacy between them. Quite a bit of jockeying goes on in the establishment of a comfortable social distance in interactions. This comfortable social distance varies between cultures; typically the Anglo-Saxons stand at a reserved distance when conversing, while the Spanish and Italian peoples stand much closer. At international conferences you can often see Americans and British people edging backwards, pursued by Italians and Spaniards trying to get to their usual degree of close-ness.

Social distance is significant. There is nothing an audience can do about their perceptions, and little a speaker can do to modify the effect. He or she must make use of these messages, to establish the social tone required.

The embattled speaker

Many speakers respond to the speaking situation in a way which both conforms to the theories of proxemics, and illustrates their subconscious attitude to their audience. These speakers head straight for a rostrum, and hide as much as their body as possible behind it. The rostrum is made to look like a crenallated gun-emplacement. Sometimes only a bald head and glasses are visible above an early perpendicular, bullet-proof, oak lectern. At other times there is only a disembodied voice, mumbling from behind a heavily engineered brass reading-lamp. I have also seen people hiding behind a boldly-printed card bearing their titles, names and qualifications like a decoration for gallantry. All these positions are just the defensive signals they look. Amusingly, it is often the most formal and senior of speakers who adopt such positions, inured by years of power and seniority, they still display their haunting insecurity by their defensive positions when speaking.

Rostrums are anathema to good speaking. It is difficult to create a friendly atmosphere if most of your body is shielded by psychological armour plating. My practice is always to push the rostrum to one side, and stand in full view of the listeners. Even standing behind a table gives a sense of social division; the friendliest solution is to stand in front of the table, with your notes slightly behind you. They can then be picked up by an easy movement, and put down again if you need your hands to add gesture to your message.

A hot presentation

What I have said so far, I hope, has convinced you that it is naive to disdain the simple, physical comfort of the environment. The practical comforts of the room, the temperature, the chairs, the heating, the lighting, the amount of fresh air, and a visible clock are all important parts of physical and psychological comfort. Research confirms that the environment of a talk profoundly affects people's unconscious attitudes to a presentation.

Griffitt showed that people meeting in a room with a temperature of 34°C, or with only four square of floor space for each person, like each other less than when they meet in a larger and cooler room. But the physical surroundings also work in another way; they have a symbolic meaning. A room which is decorated in red and yellow seems to suggest a warm emotional mood; placing some of the people higher than the rest, using a high table for instance, suggests dominance; a room with a concrete floor, battered furniture and a bare light bulb has unpleasant associations of prison or an interrogation.[1]

John Mitchell warns us that: 'the audience holds the man standing up in front personally responsible for any discomfort.'[2] This disturbing responsibility is easily discharged, though. A few minutes inspection of the room before the presentation will give you a chance to check lighting, heating, and seating. Ventilation is difficult to gauge without all those bodies present. But check which windows will open, and notice if any of them open onto noisy roads or factory buildings. Remember that moving air, even if it is no cooler, may have an arousing effect on people, just because it provides a marginal extra stimulation. It is a simple point, but like so many simple points in the preparation for speaking it can make a major difference.

The social interaction when giving a talk is so delicate that the smallest problem may spoil it. It is unthinkable that the first time anyone goes into a room should be when they are starting to speak in it. Can you imagine taking a new friend home after an evening out, and it being the first time you have walked into the room in your life? If we prepare everyday social interactions, we should prepare a major career event. Yet, amazingly, it is common for a speaker not to be shown the room he is to speak in.

When you are speaking, try to have a spare half hour before the talk, and ask to see the room you will be using. It is much easier to prepare effectively if you have a clear picture of any problems. A quick walk round with a mental check list of 'chair/heat/light/vent', can save much difficulty later. Never allow yourself to be surprised by the room you are speaking in. My practice is always to ask to see the room before any reception or general introductions, soon after I arrive at the organization where I am to speak. Mentally running through the check list often uncovers problems which are easy to fix before the talk, but would cause major disruption during the talk.

Ventilation can be dealt with during the talk, providing your presentation is not rendered ridiculous by a long tussle with unopenable windows, with much clambering on tables, sweating and straining, and effusive apologies from embarrassed organizers and distracted speakers. Remember that audiences are generally colder than the speaker. Conversely, a speaker rarely finds a room sleepy; an audience rarely does not. You must be alert to the signs of doziness in the audience, and do something to improve ventilation if it is caused by stuffiness. One solution is to point politely to someone, and ask them to open a window. It demonstrates concern for the audience's comfort at the same time as showing authority, and boosting your own confidence. The person asked never refuses, and the speaker is reassured that the audience is willing and kindly.

You should also consider what arrangements there are for refreshments, and coffee? Do they know whether to expect coffee, and if so when? If possible, ask the organizers to avoid the waitresses banging into the room and starting to pour the coffee during the closing minutes of your talk. It is also kinder to you and the audience to ask the organizers to try to prevent the tantalizing smell of fresh coffee wafting through the door a quarter of an hour before your scheduled finishing time, especially if yours is the first session after a particularly convivial evening.

Rows and rows of chairs

In making the physical arrangements for the talk, you must consider the audience's position, as well as your own. The most important question, here, is how large is the audience going to be? Different arrangements are appropriate for different sizes of groups, and types of presentation. The first question to decide is whether you are going to arrange the room as a circle or a row? I said 'you are going to arrange the room' deliberately; few rooms are arranged in the best possible way when you first walk into them. Typically, a room will be arranged with serried ranks of chairs in a block formation, just like the village meeting hall. There is a reason for this – it is the only arrangement which will get the absolute maximum number of people in a room. Organizations usually have the maximum number of chairs they can squeeze into a room, which can then be recorded as a '60 seater', for scheduling.

If you are expecting an audience of 60 people, there is little you can do but accept the sardine can arrangement. But it is more common to find your talk put into a room which is larger than the expected audience. If you are speaking in your own organization, you can probably make a choice among rooms. If you have spare space, consider different arrangements of the room. If the group is only fifteen or so, try to make the tables into a circle in the centre of the room, stacking the spare chairs in the corners. If the audience is bigger than twenty, you can still improve the psychological impact of the room by moving the chairs into a broad semi-circle, centred on the speaker's table. Chairs are best placed in a 'U' or a 'C' shape, with the speaker at the open end. But make sure that the sides of the 'U' are not long and straight, so that the effect of an army parade is not produced. If you put out one circle of chairs, people are forced to sit with no one in front of them. The effect is that everyone feels involved in the presentation, and no one can doze on the back row. If you stand in front of the speaker's table, they can also see you, and feel a link with you as a person as well as a speaker. This arrangement creates a warmer sense of friendliness, and the open space in the semi-circle removes the crowding and restriction that comes from having a chair in front of you. People can stretch their legs, and they feel able to breathe. These arrangements are illustrated in Fig. 10.1.

Many speakers seem to be afraid to move chairs and tables in a room; as if the dust of ages were sacrosanct. But most organizers will be happy to help in a five minute removal exercise, if you express your

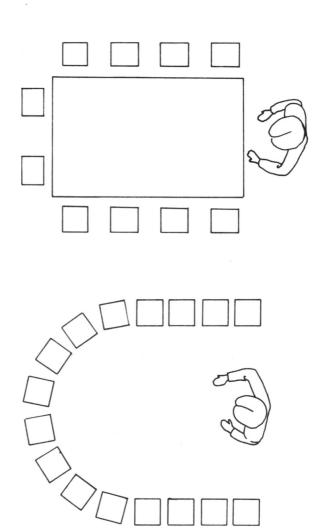

Fig. 10.1 Seating arrangements for groups up to 30

intentions and reasons clearly enough. One of my colleagues always moves the furniture as his first action when he has to speak in a new room. He creates comfortable, intimate spaces in the most forbidding halls; the reward is a level of attention and interest in his talks which I have seen few people match. Furniture removals are good tactics. Unless the room is perfectly arranged, or the chairs screwed to the floor, I strongly recommend it as a normal act before any talk in a strange room. If more speakers did it as a matter of course, there might be fewer rooms permanently arranged in staid, cramped, and unfriendly rows.

When the group is over 30, it is probably too large to put in a semi-circle, and you must then use rows. The favoured solution, after the single semi circle, is a double circle. This layout will accommodate up to forty people, without anyone feeling isolated from the focus of attention, which should be the speaker, not the back of the head of the person in front. You can make sure that everyone can see by staggering the second row, so that people look between the shoulders of the people in front. The point is so simple that you would think everyone would do it. But either ignorance or inertia makes such a simple arrangement rare. Figure 10.2 shows this arrangement.

If your audience is more than 40 people, and the room is fairly small, you may have to accept the arrangement of chairs in rows. I have said enough, I hope, to make it clear that I think this a last resort. It always results in an uncomfortably crowded and unfriendly sense of space in a room. The back row becomes a problem area, because serried ranks of chairs are daunting. Secondly, the back row look at the speaker through the other rows' heads and backs. Also, the front row sit with people staring at their backs which is uncomfortable, because they never know when, and if, they are being watched. So everyone hunches over their notes, afraid what those behind are thinking.

A comfortable arrangement

It has been shown that most people talk easily to those sitting opposite them and those they can see, rather than people sitting beside them. It is even more difficult to talk to someone the other side of your neighbour in a row. Not only is it physically a strain to lean round to see them, it is not possible to see or use a full range of non-verbal signals. Because of this, people feel a sense of community with the person they are facing, sitting in a 'C' or 'U'. Considering the audience' comfort does not end with arranging the room. You need to ask yourself how old the audience is. What they will find comfortable will be affected by how much comfort they are used to. Hard upright

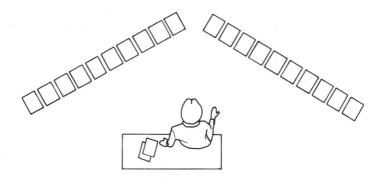

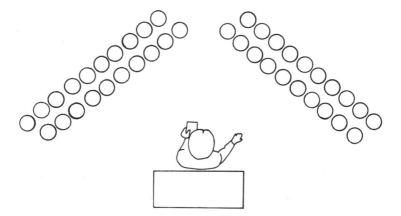

Fig. 10.2 Seating arrangements for larger groups

chairs will not suit an audience of older, senior people, although they may be fine for an audience of young people. Equally, low arm chairs may make it difficult for people to listen to a detailed technical talk, although they may be fine for an evening presentation about a hobby or personal interest. Often, you will have little choice in the chairs to be used, but if you can choose, make sensible use of it. The chairs, like the arrangement of the room, are all part of the effects you can manipulate to create a perfect presentation. Like a virtuoso conductor, you can get the best out of your score by careful preparation.

The ways in which people can be uncomfortable during a talk are many; sitting listening to someone else for long periods is an activity which attracts discomfort. The discomforts of being hot, stuffy, or doomed to a hard chair grow as the talk progresses. But the final type of discomfort, psychological discomfort, decreases as the audience get used to their situation. Usually, psychological discomfort never becomes conscious, and is therefore rarely blamed for unpleasant associations. But it is real, none the less.

What causes psychological discomfort? There are two main causes; being in an alien place, and being too close to too many people. An alien place is not something out of science fiction; it is simply anywhere which is not felt psychologically to be home territory. If the talk is being given in neutral space, public rooms, or a familiar lecture room, the audience will soon relax and feel at home. But if a presentation is being given in someone's private office it will take longer for people to cease to feel subtly threatened. The owner of the office is probably the only person who is completely relaxed. Everyone else has a faint sense of trespassing, and that slight level of added arousal everyone feels when on alien ground. The speaker must be aware of these subtle feelings, and since he is a neutral intermediary, make every attempt to be friendly and welcoming, and demonstrate that he is relaxed and at home himself.

The size of the room in relation to the number of people in it also affects psychological comfort. Too many people in a tiny room is uncomfortable, not only because it is stuffy, but also because people cannot achieve a natural spacing. A hidden defensiveness and hostility develops when people are not able to leave adequate space between each other. Equally disturbing, is too few people in a hall like an empty barn. A natural spacing of people, in a neutral space, is the best condition for a talk. Perhaps the worse condition is a small, crowded presentation in the speaker's own office. People are un-comfortably close, and just because it is the speakers own room, the audience feel they are being interviewed, rather than talked to. If you have to give a small presentation, it is essential to try to find some other room, even the office of a colleague is better than your own office. You can find a slightly bigger room, to give plausibility to your change of venue, although the real reason is psychological.

Psychological discomfort is the subtlest of the various types of discomfort caused by thoughtless physical arrangements for a talk. But it is by no means the least important. Indeed, just because it is difficult for the listeners to identify the cause of their discomfort, they are more likely to attribute their misery to a bad talk, than to bad arrangements. By being aware of these factors, you can increase the chance of a good reception from your listeners. Fine arrangements will not make up for badly prepared material, presented in a boring way, and running on far too long. But bad arrangements can undermine careful preparation, and make it a little more difficult to hold the audience's attention.

Practise with a friend

All this advice about the physical arrangements of the room has been very general. And it is difficult to give exact rules which will fit every situation. Therefore your best method of learning is to gain experience from watching others, and listening to what is said to you. A useful aid to preparation, if you have the chance, is to listen critically to someone else giving a talk in the same room. It will give you a vivid picture of the advantages and disadvantages of the room. You will learn from the other speaker's mistakes, and you will find out exactly what an audience feels like in that room. Incidentally, the best place to sit is the back corner of the room, because you will then understand how important it is for the speaker to be aware of the whole audience. But if you can go to several talks, try different seating positions to see different effects. You may pick up some useful tips about the audience's view of the speaker!

Perhaps the the final key to a well prepared talk is to practise with a friend. Use the actual room you will be using on the day, and give a complete presentation. Listen carefully to what he or she says, even if you disagree, the judgement of an outside observer is always more accurate than the wishful thinking of the speaker himself. I have already said that we are very poor at judging what we sound like to others. It is equally difficult to assess what it feels like to sit and listen to ourselves. The friend is your only objective evidence about what works well, and what needs to be improved. You can also enlist a second opinion about spacing of chairs, heating, lighting, and noise.

You will certainly be asked to speak again, some other time in your life, so use the friend to add to your stock of experience, and knowledge about yourself, speaking in general, and the effect of physical space on the audience. If your friend is in the audience on the day, ask him or her to make constructive criticisms. Prompt him or her with alternate questions about good and bad points. If you perform the service of objective listener for someone else, remember that what you say may carry a lot of weight. Few people hear factual criticism of their presentations, for listeners are usually too polite to say anything. The speaker may have lived in a fantasy world, and be raw and sensitive about his actual performance. I have found from experience that it is best to mix positive and negative points. Try to boost confidence by pointing out first the attractive features of the speaker's technique. Balance this by noting what can be improved. Even very experienced speakers can improve; but a new and inexperienced speaker needs most of all a boost in his confidence, and praise is the best form of criticism to achieve this result. So encourage your friend to make the praise real and credible by also noting what needs to be improved, but to balance the apportionment of praise and blame. The aim is to improve your speaking skills, not to indulge in a self confidence obstacle course!

Notes to chapter ten

1. Griffitt, W., and Veitch, R., Ten Days in a Fall-out Shelter, *Sociometry*, Vol.37 (1974), pp.163–173.
2. Mitchell, John, *A Handbook of Technical Communication* (Wadsworth, 1962), p.186.

SUMMARY SHEET

Chapter ten – Environment

Audiences are sensitive to physical discomfort.

Rearrange the room to help them listen.

Audiences blame the speaker for every discomfort.

Proxemics is the study of how close people place themselves when communicating.

Some speakers shelter behind a rostrum.

If the room is hot and stuffy, people will dislike you.

Take a few minutes before the talk to check lighting, and ventilation.

Rows of chairs are intimidating. They can usually be moved into a circle, 'C', or 'U' shape.

People in the back rows look over others' heads; people in front feel watched.

Psychological discomfort is caused by intruding into someone's private space.

People talk more easily to those opposite them.

Get a friend to listen to you, and comment on the arrangement of the room.

Listen critically to other speakers in the same room; you will discover what it feels like to be in the audience.

11

Visual aids

What is the use of visual aids?

Most experienced speakers use visual aids. Why? The short answer is that they make the talk more interesting for the audience, and they do this for a number of reasons. Let me suggest seven reasons, before giving practical advice about the common types of aid, their virtues and vices, and the pitfalls for the inexperienced. Because there are many different aids which speakers use, and because each of these aids has its own technique, with many small details of management, this chapter will be rather different from the rest of the book. It would be tedious to read through all the details at one go. Indeed, if you are reading this book from front to back, I recommend you to read only the first few pages of this chapter, before moving on to Chapter Twelve. Skim through the detailed advice on individual aids, but please don't try to read it all. It is not meant for light reading. I suggest that you come back to the advice about a specific visual aid, only when you have to give a talk, and will be using that particular aid. In this way you will have the motivation to read and absorb the detailed points of technique; you will also have them fresh in your mind when giving the talk.

What are the seven reasons why visual aids improve presentations? The first is that visual aids get *attention*. People are naturally more interested in things and pictures than in abstract words. Even adults look through books for pictures. So visual aids have a greater impact than words alone. If you are an observant listener at other people's presentations, you will notice that there is always a stir of interest in the audience when a visual aid is shown. When there is something to look at, people wake up.

The second reason follows from the first. I have already said that the best way to keep an audience awake and attentive is to provide variety. Visual aids give *variety*. Audiences can only listen for about ten minutes without having a micro-sleep. Most people mentally wander off at irregular intervals for a little day-dream: all sorts of daily concerns absorb their attention for a few seconds. So, the normal span of attention is very limited. Visual aids are the most convenient way to break up the talk into smaller chunks, with a fresh picture or object every five minutes or so. The effect is that before the natural span of attention is exhausted, there is a fresh interest to arouse it again.

Notice how you yourself respond to talks with, and without, visual aids. I think you will discover that you find it much easier to keep awake, and pay continuous attention to what is being said, when there are aids than when there are not. Visual aids, then, provide relief and change. If possible, introduce a new aid every five to ten minutes. It prevents monotony, and provides an injection of refreshment to the flagging concentration of the listeners. Lloyd (1968) showed that the audience's level of arousal falls steadily throughout a talk, until just before the end when it rises in anticipation.[1] It is the speaker's job to try to keep this already wilting arousal as high as possible, and a variety of stimuli is the only way. Punctuating the talk with illustrations and other aids is a way of keeping arousal at a high level throughout the talk.

The most important reason of all, though, is the third one. Visual aids help *memory*. To begin with, they are the best way of experiencing facts. Psychologists tell us that the order of effectiveness for learning and remembering facts is:

> Seeing the real object or event
> Seeing a picture
> Being told about it
> Reading about it.

Another reason why visual aids help retention is that they improve the audience's sense of the structure of a talk. In writing we can go back and check what has already been mentioned, but in speaking we can not. Even the simplest visual aid can provide a grasp of the structure and direction of the argument which will help the listeners to understand, and remember. Something as simple as keywords on the board reinforce, confirm, consolidate, and focus attention.

There is evidence from studies of memory, that memory works better through sight than sound (for example, we remember a face before a name). Experts in learning have demonstrated that information which comes into the mind through several different senses is more likely to be retained than information which comes from one sense alone. Visual aids use more than one sense to fix the message, and this is why a talk which uses aids is more memorable than a talk which does not. It is also why a talk with a lot of technical details *must* use visual aids, since people simply cannot absorb and understand many facts and figures without aids to help them.

The next reason why visual aids help the audience is that they *save time*; visual perception is faster and surer than verbal description. A complex shape, flow-path or relationship is difficult to describe in words, but can be quickly and memorably shown in a picture. Indeed, there are some things which it would be virtually impossible to describe in words, which can be easily shown in a picture. It is interesting that graphical presentation of information is also becoming a popular feature of business computers. Computers can easily present page after page of figures, but human beings find them difficult to grasp. Graphic presentations in bar charts, pie charts, and graphs are much easier to grasp. The phenomenal success of new software, such as the *Lotus 1-2-3* package[2] is a sign of how convenient, and time saving, graphic presentation is. Visual aids are graphics for speaking. If you have the right equipment, incidentally, it is possible to turn the graphs produced by computers into slides or view-foils for projection during a talk.

A fifth reason why visual aids are helpful is one which is often not thought about. Visual aids provide the *same image* for all the audience. The word 'house' produces a different image in everyone's mind, but a picture of a particular house does not. Similarly, if you tell your audience that the chemical plan for the old process is in a 'dirty and untidy' state, different members of the audience will have different pictures of what you mean. But if you can show a slide of the old works, you will be sure that everyone has the same information. The same is true of facts and figures. If you tell your listeners that the growth rate of the company has 'speeded up', they may have a variety of different ideas of what this means. If you show them a graph, they will all have the same picture. Visual aids make sure that everyone has the same thing in mind.

Some information *can only* be presented visually, for example a complex shape, the rate of change of a curve, complex statistics, or mathematics. Geniuses who can do rapid mental calculations seem to visualize the figures in order to manipulate them. They have a sort of inner pencil and paper on which they see the calculations. Even geniuses need to visualize things; for most of us, there are some things which we can only understand as a shape. Katona (1940) showed that a method of teaching a manual skill which used only words was less effective than a demonstration.[3]

As I mentioned earlier, the Romans understood very well that our mental processes rely on our sense of shape, and place, to organize themselves. They had a popular system of remembering the points they wanted to make in a talk by associating each point with a corner of a familiar room. One point would be associated with the doorway, the next with the left hand corner (where there might always be a vase), and so on. It is from this system of memory that we get our phrases 'in the first place . . .' and 'in the second place . . .'. If simple organization is best remembered in a visual way, think how much more difficult it would be to understand complex spatial and geometrical relationships without visual aids. But even if you have good visual aids don't overload your audience's memory with too many figures in a presentation; you are not holding an exhibition, but getting over an impression. Two final advantages; visual aids also *help the speaker*. They help with stage fright by giving the speaker something to do. They use up excess nervous energy which would otherwise be frittered away destructively in fright and apprehension. They do this in two ways. In the first place, they give the feeling of being well prepared, and therefore safe. In the second place, they help the hands and body to move naturally. Instead of trembling, there is now something for them to do in setting up, moving and displaying the visual aid. But nerves can still show themselves in strange ways, and you must beware of showing your apprehension by producing a vast number of over complex and unnecessary aids. You must also beware of the tendency to handle, or point to, the aids in peculiar ways. Nervousness often shows itself in rhythmical actions: trembling is just one of them. Nervous speakers can be seen rhythmically putting up and removing pictures, or stabbing regularly at the projection screen with a pointer. Such mannered use of aids will disturb the audience's concentration. Remember to leave the aid strictly alone, while you talk about it, and only touch it when you are ready to remove it. The handling of aids should be done firmly, and briskly. If you do this, they will help to control your nerves.

The final advantage to the speaker is that visual aids can help the speaker's *memory*. They can be a list of key points, which help to remind the speaker what comes next without needing to refer to his notes. He or she can simply take the points in order from the visual aid. In this way, the speaker doesn't have to keep track of two sets of information – the notes and the aids. He or she can also concentrate on the audience, and making sure that the aids are presented clearly, with the confidence that he or she won't forget to say anything, and won't lose the place, because the next aid will remind him or her of what comes next. Thus, aids help to ensure the talk has a structure, as well as helping to ensure that the audience can see that structure.

The disadvantages

If so much can be said for visual aids, why does anyone ever speak without them? Or why speak at all? Why not just hold an exhibition? What I have said so far might suggest that aids do all the things which are necessary to keep an audience interested, and communicate information. Indeed, I have seen talks where there were so many aids, and so little to say about them, that the speaker was doing little more than hold an exhibition, occasionally pointing out something which was obvious anyway.

This is clearly wrong, and we must think carefully about the limitations of visual aids, as well as their virtues. The reason why aids cannot do everything by themselves is that personal contact between the sender and the receiver of the message is an important part of communication. It is also true that people are much more interesting than things. We are visual aids in ourselves. Our facial expressions, gestures, tone of voice, movements, and the atmosphere of enthusiasm we (hopefully) generate all illustrate the verbal message. The best aids of all are gesture, appearance and tone of voice, because they are dynamic, moving aids, and they are synchronized with the rhythm of spoken message. Gesture is integrated with words, and conveys information in a familiar way. That is, of course, the main reason why a talk is preferred as a way of gathering information to books, recorded tapes, sets of pictures and other non-living media. One of my colleagues is a man of such presence that his appearance is a dramatic performance in itself. He is his own visual aid, and he is probably the best speaker I know.

A second answer is that visual aids by themselves don't tell the story. We need a guide to an exhibition; captions and legends need explaining, and important details need pointing out in a logical order. There is usually no obvious order in which to read a picture, but it may be necessary to have noticed some information first before being able fully to understand the point of some other information. Words help here. Few of us are trained to 'read' pictures efficiently. Davis and Sinha (1950) showed that 'a story told before seeing a picture may affect how the picture is remembered.'[4] What we see depends, very much, on what we have pointed out to us. Aids without a talk would be even less use than a talk without aids. They are mutually supporting ways of presenting information.

A further reason why visual aids cannot solve all the problems of presentation by themselves is that some concepts cannot be made visual. Ideas, such as entropy, mass, or 'importance', are purely mental concepts, and I cannot easily imagine what a *picture* of entropy would look like. Many more abstract and general concepts are of this kind. Aids can still be used to support the talk if it is about such subjects, but they will be verbal aids, with keywords and phrases on them. Incidentally, if you try to present large stretches of text, you might as well write a book as give a talk. Aids with too much text in them are tedious for an audience. Only talks which have to do mainly with objects and things lend themselves naturally to many visual aids, though admittedly this probably includes the majority of technical presentations. But in most cases a few aids, in a supporting role to the spoken message, is the ideal.

A final important point against the excessive use of visual aids – too many visual aids can *distract*. Too many visual aids can become visual handicaps, by making the presentation disjointed, and making people forget the structure of the speech while looking at the pictures. I have seen presentations where what was being said was completely outweighed by a mass of pictures. The result was boredom and confusion for the audience. Aids can distract the speaker too. If you have too many aids, it becomes difficult to control them, and too much time may be spent changing slides, or turning flip-charts, and too little time given to discussing the points they make. Speakers can be over-whelmed by too many aids, as well as audiences. Often, one of the most difficult decisions in preparing a talk is cutting down the number of aids to a reasonable proportion. I usually find that about seven overhead projector transparencies will cover a 50 minute talk. One aid, even if it only contains half a dozen key words, is enough for five to ten minutes. Any more will simply distract both audience and speaker.

Ten points to watch

My advice on the use of visual aids can be concentrated into ten points, which are basic to the good use of visual aids of every kind.

They are general warnings which apply to all aids, and they are drawn from watching about 1,000 presentations employing aids, and trying to pinpoint what worked, what failed, and why. A good way of ensuring that your use of visual aids is as effective as possible is to try to check through all these ten points one by one, honestly asking yourself if you have complied with their advice.

1. Appropriateness

Only use a visual aid if it *improves* the spoken word; always keep the real purpose of the talk in mind. Mentally test each aid, and think whether it is the best way of making a particular point. As with all communication, you must make the decision on appropriateness by considering the audience and their needs, before your own convenience as a speaker. Intellectual audiences find too many aids distracting; they slow down the rate of delivery of the information. Audiences of moderate to low ability, with small prior knowledge of the subject, and little experience in that field, need more visualizations. A good number of aids reinforce the ideas and information by repetition, and give the listener much needed time to absorb new facts. Ask yourself: 'what will this particular audience expect in this particular context? Unusual, unfamiliar, or childish aids can spoil concentration. What will be the preferred mode of communication for this audience?'

Do not let what is available, and especially what is merely fashionable, dictate the aids you use. Decide which aids to use on grounds of content and purpose; then look for aids which will do what is required. Get the aim and purpose of the talk clear before choosing the aids. Try to do a 'systems analysis' of the best presentation path, do not mangle and squeeze the content to suit the convenient aid. Try to use several different aids. Avoid thinking of your presentation as a 'slide-talk' or a 'chalk-talk', for if you do, you will distort the information you have to present for the sake of the aids. The result will certainly be inefficient, and may even be confusing for the audience.

In general, the most important thing to do before making any decisions about individual aids, or doing any work on preparing the aids themselves, is to analyse your strategy: 'What am I and my aids going to do that a talk without aids could not do more quickly, more efficiently, and more cheaply for this audience? Am I simply saving my time and effort or am I genuinely saving the time and effort of the receiver and user of the message?'. Remember, in making this mental calculation, that the time of an audience of 20 or more is very expensive in total man hours. It is rarely cost effective to save much of the speaker's time at the expense of the audience's, even if the speaker is the managing director himself. So think out an effective strategy, and be prodigal in the expenditure of your own time where it will save the audience's time. You must make sure, too, that the visual aids were used, not merely displayed; in other words, you should have at least one substantial point to make when you show each aid.

2. Clarity

The next thing you must consider is the overall clarity of the aid itself. The point of the visual aid should be immediately clear, and the aid must be easily seen. It should not be necessary to spend a great deal of time explaining the aid; its meaning should be obvious. Also you should not need to warn the audience to disregard irrelevant parts of an aid. If so, you must draw a new one. Another point which often affects clarity: while two or three different colours help to give the aid impact, too many colours can be confusing. Not only must the aid be clear and obvious in itself, you should also make sure that it can be seen. It may look fine from where you stand – close up – but can everyone see it? Can the audience see the aid from the back of the room?

There is only one way to answer this: before the talk, go into the room you are going to use, put up one of your visual aids, and go to the back of the room. You should sit down in the chair which is furthest from the aid, and see how much of it you can see. Remember that during the actual talk, the row in front may also be full of people, and it will be even less easy to decipher a remote visual aid. Check to see if the board is shiny, or if there is an obstruction in the way. Check that there are no reflections from windows or lights which will interfere with the visibility of the aid. You must also make sure that the aid is big enough. It is surprising how much difference distance makes. An aid which may seem huge to you, standing next to it, can be almost illegible from the back of a large hall. If you have to speak in a big room, this check for visibility is especially important. It is quite impossible to guess what the aids will look like from the back of the room; you have to go and see. Believe me, I've tried it often enough, and each time I am surprised! If the room is too large for the aid, the only solution is to prepare a larger aid.

3. Simplicity

Many of the aids I have seen have failed because they are too complicated. A maze of lines, and a book-full of text, will only confuse the audience. The very minimum of simple childish diagrams, and single words, is quite enough. Visual aids are always clearer if they are simpler. As a general rule use only the minimum summary of the facts and points you are discussing. The aid is meant to reinforce your message – no more. It is not meant to be the message itself. Thus, the audience find the visual aid easier to understand if it contains outlines, not fully detailed pictures.

Too much detail will confuse them; for rapid understanding keep detail to a minimum. If there is a lot of redundant or unnecessary information, they will wonder what all the detail means, and if and when you are going to explain the meaning of the schematic representation of some tap or dial you have put in for realism, not for information. Even worse, the audience may not be listening to what you are explaining, because they are curiously examining the irrelevant details on the visual aid. They may also mistake which exact point or line you are talking about if there are too many of them. If you have a lot of information to present, the sensible solution is to use several simple aids, rather than one complex one.

Even if the aid does comply with this advice and is an exemplar of clarity and simplicity, you still need to remember to leave each item for the audience to read and digest before talking about it. Most speakers wildly overestimate the speed at which a listener gets his mind around unfamiliar visual material. They also tend to over-estimate the listener's capacity. Only use three or four lines of text at the most – a visual aid is not a reading text. A 'blank wall of prose'[5] should be avoided. A carefully designed, simple visual aid should be substituted for a photocopy of an original text, which is too easy to make, and almost useless as a visual aid. A visual aid should almost never contain even complete sentences: condensed, telegraphic notes, or keywords will do – the talk itself will explain in full. The visual aid is simply a skeletal reinforcement for the talk. Above all, avoid showing a full minute's worth of reading matter for a few seconds only.

4. Are they large enough?

I have already pointed out that the aid must be legible from the farthest seat in the room. The point is so important, and so often forgotten, that it is worth repeating. The aid must have clear, large, bold and uncrowded drawings and lettering, which can be clearly seen from everywhere in the room. Remember, some older members of the audience may have poor eyesight; some may have brought the wrong glasses, some may just not bother unless the letters are clear and easy to read. A rule of thumb is this; stand at the back of the room and look through one eye. If the aid is eclipsed by an extended fist it is too small. Prepare a bigger one.

One point which is nearly always forgotten: if the size of the drawing or writing is increased, the thickness of the line used must increase also. Remember that, for the back row, a label or caption which cannot be seen is worse than no label or caption at all. It is surprising how often a flip-chart has enormous letters with spidery thin lines, or how often a nice drawing has a tiny, inscrutable caption. The effect of such simple mistakes of technique can be strangely catastrophic. Speakers feel, quite naturally, that if they have given an inspired presentation, fiddly details like the size and legibility of their visual aids should not count for too much. Not so. If the visual aids are not visible, any talk, good or bad, will be ruined. These points of technique are simple enough; getting them right is easy, but getting them wrong is disastrous. It is, many speakers reflect, an unfair world; but a little thought, and perhaps checking through this list of points just before you speak, may dramatically improve the effect of your talk.

5. Are they distracting?

The point of a visual aid is that it should aid you, not replace you. Yet visual aids can sabotage the speaker if they are ill-planned, poorly timed, and irrelevant. Aids tend to shift the attention from speaker to the picture, and so they can easily break up the unity and continuity of a talk. The presentation can too easily lose its force and momentum, if too much time is spent fiddling and adjusting projectors and flip-charts. Often when a first aid is introduced the speaker feels distracted himself. He loses the rhythm of his presentation, and starts to mumble. He finds that he has lost the thread of his talk, mislaid his notes or lost his place in them, and he lets the pace of the presentation sag.

In some cases, such as teaching or demonstrations, the aid is central to the presentation, and the talking is peripheral. But all visual aids still require a verbal framework. They need explanation, and linking into the conceptual framework of the listener. Research shows that a verbal framework is essential for visual comprehension,[6] and where no explanation is offered for a picture many people fail to see things, or think they have seen something which is not really there. The caption of a picture is an important part of how much and what is remembered. Similarly with visual aids for a talk; the flow of the argument will flounder, or become distracted and diverted unless, when the aid is first shown, the speaker clearly and succinctly describes what its purpose is, and what the audience should be looking at. The simple action of a verbal summary will help the speaker too, for it focuses his attention, as well as the audience's, on what he meant the aid to do for the talk.

The simple fact is that people need telling what to look for in a visual aid; their eyes need directing, and their attention controlling. They see what they are told to see (sometimes even when it isn't there). Unless you use the ability to direct their perceptions they will let their minds wander, and become distracted. When preparing aids, think carefully about the minimum of clear information you need. When presenting the aid, concentrate the audience's attention on the point the aid is trying to make.

6. Cover them up

Visual aids distract an audience if they are left on view when the talk has moved on to a different topic. One clear way of signalling that the topic has been closed, and a new one started, is to cover up the aid you have been using. This action deprives the audience of something to stare at, and their attention will swing to the speaker. It is an ideal moment to make an important point; you have roused their attention by the change in what the are looking at, and for a few moments you will have their full concentration.

I usually use these moments to sum up the point I am finishing, point out that the talk is now moving onto a new topic, and state the main point of this topic. Then, when the audience's attention is beginning to flag again, uncovering the new aid will revive it. It is a useful technique; if you have ever seen anyone using it, you will have noticed that the covering and uncovering, or switching on and off of an overhead projector, is not distracting. In fact, it nicely punctuates the presentation. The effect is like a conductor, using gestures to shape the information.

The valuable effect of covering something up is just as noticeable before a visual aid is used as it is after it is finished with. Covering-up aids before they are used lends an air of expectation; it keeps the audience in mild suspense, and reinforces their interest. When the aid is finally uncovered, there is a sense of satisfaction which further arouses their attention. But don't tease the audience; it is best to keep the aid right out of sight until the moment it is referred to.

A practical example of the technique of covering and uncovering aids is the flip-chart. In a pad of flip-charts, leave a blank sheet between each chart. It is then possible to turn the chart you have finished to the back of the stand, leaving a blank sheet while you sum up, and announce the next topic. Turning the blank sheet over to follow the previous sheet then reveals the new chart. The technique is simple; but its effect is out of all proportion to its simplicity. For all the reasons I have already given, it helps the listeners to grasp the information. It also gives an impression of thoughtful efficiency, and careful preparation, which will boost their image of the speaker almost as much as the content of the talk itself.

7. Leave them long enough

Most speakers race through their visual aids. The audience is given a quick flash, and just as they are getting their eyes in focus, and their minds ready to absorb the new information, it is whipped away. If this happens several times, people give up trying, and often give up listening as well; visual aids take more time for a (possibly already sleepy) audience to focus on, than a nervous and jumpy speaker realizes. It is most important to leave the visual aid on display long enough for full comprehension, and careful, detailed reading by the audience. They will understand and read much more slowly than the speaker because they are new to the information. You are also keyed up, and full of adrenaline, whereas they are relaxed and sleepy. Visual aids need ten to fifteen seconds silence, when they are first shown, for the audience to look them over. This may not sound very long but if you look at your watch for that length of time, you will find that it is a substantial block of silence. Yet many speakers leave at most two or three seconds, or start explaining breathlessly, with no pause at all.

Even simple pictures need quite a time to understand; slides need a full 15 seconds before you start pointing the audience's attention to particular details. If you are showing textual material, or graphics, try to leave a silence long enough for people to read right through – they will not notice it as a gap in the presentation. Read through any text material to yourself slowly, *twice*, before judging that the audience have had long enough to absorb it, and continuing the talk. And don't read the slide *for* the audience – it's insulting and repetitive. The audience will not notice the silence, because they are reading and absorbing. Their minds are still occupied, and their ears will welcome a change from the speaker's voice.

8. Don't talk to them

The majority of the inexperienced or nervous speakers I have watched turn to look at the visual aid as soon as it is displayed, and never turn back to the audience. They end up, with their back to their listeners, casting occasional nervous glances over their shoulder to make sure the audience is still there. Contact with the audience is lost, and the speaker's voice dwindles to a mumble. Visual aids are comfortingly familiar to you as speaker, and are sometimes used almost like a baby's dummy. They are your own work, they are familiar, and above all they don't look at you, talk back, argue, move, or yawn. And they don't have many pairs of enquiring eyes! Inexperienced speakers tend to talk to their aids, rather than to their audience. But turning away from the audience, and talking to a flip-chart, board or screen breaks contact between speaker and audience. If the speaker has his or her back fully or partially turned, his voice will be muffled, and, as I explained in an earlier chapter, the auto-volume adjuster will reduce the volume of the voice because he or she is now talking to a nearby object.

Chose your standing position carefully so you do not have to turn your back to the audience to point to the aid. Glance at the aid, but centre your attention on the audience, rather than the other way round; some speakers make furtive glances at the audience while fixing their visual aid with a long, admiring stare. A position well clear of the board or screen so that the audience can easily see it will be more natural if a pointer is used. Some years ago a portable pointer appeared on the market which looked like a metal ball-point pen until it was opened up like a telescopic car-aerial, with a red knob on the end. Rising executives always had one in their top pocket, and would erect and collapse them with great aplomb. They are, in fact, useful aids, if you can keep them still, and put them away when they are not being used.

9. Don't hand out at random

Unless you have a copy for everyone, don't pass anything round, large or small. It breaks up the audience's attention. The object attracts attention to itself like a magnet in a circle of iron filings. People get excited when their turn approaches, and cease listening to the speaker when they finally get their hands on it. Several objects are even worse, since they mean that each individual is thinking of a different thing at any one time, and the speaker cannot possibly be talking to the whole audience.

I have seen presentations where a collection of photographs have been passed round. It was fascinating to watch everyone with their head bowed over one or other photograph, totally unaware of what was going on around them. The speaker continued to talk, but no one was listening. They were all far to anxious to hang on to what they had got, or nudge their neighbour to hurry up with the next one, or whisper to their other neighbour about the one they were just passing on. The talk, as a talk, fell apart. The moral is that it is impossible to control the attention of a group of people unless they are all doing the same thing. I have more to say on this topic later in the chapter when dealing with handouts, but as a general principle you must recognize that nothing can be given to the audience during the talk unless everyone has the same thing, and their attention is explicitly controlled. Otherwise chaos soon results. Random distribution kills communal attention.

10. Record everything in your notes

The final general point is again a simple one, yet it is one which is often forgotten. I have even seen a talk where the speaker introduced a new picture with an embarrassed pause, followed by the remark, 'I don't know what I put *that* one in for'. Nothing shows poor preparation more glaringly than visual aids. They are an extra dimension to the talk, and unless they are orderly, and controlled, the talk will soon collapse into an embarrassing chaos. Before you start a talk with visual aids, essential preparation is to make sure that every aid is in order and is numbered and clearly marked in your notes.

Slides must be checked through each time to make sure they are correct. I had a senior colleague who delighted in putting one of my slides up-side-down while I was having coffee, just so he could teach his junior colleague to check more carefully. The trick always produced the effect he wanted; the audience were always delighted, and I was always disconcerted. But I learned. I still run through the complete batch of slides just before I start, every time. It is also important to make sure the slides fit smoothly into the talk. You must record when the slide will be shown, and the exact point you want it to make, in your notes. The commentary must also be carefully tailored to the aid. Well prepared visual material is often ruined by the incoherent burbling of the speaker in the background. Don't forget which points to make, or make them in a chaotic order, as you notice them on the aid, apologetically murmuring, 'By the way . . . ' and, 'Incidentally . . .' and, 'Something I forgot to mention earlier . . .' It is, of course, a simple matter of preparation: visual aids merely make poor preparation glaringly obvious.

The many types of aids

I said at the beginning of the chapter that there was a lot of detailed advice to give about individual aids. This section deals with the main groups of aids in turn, and makes a series of points, some small, some large, but all important because the techniques of management affect the impression your talk leaves. Because there is a lot of detail, I have set out the points as separate notes. The rest of the chapter is not meant for steady reading: it is for skimming and reference, though you should look carefully at the advice about any type of aid you use regularly, or are intending to use for your next talk.

Handouts

- Never hand out anything until the time comes. Get help from an assistant, or from the audience, to speed the process of handing out. When you have asked someone to distribute a sheet of paper to the back rows, he or she feels involved in the talk, and will listen carefully to the rest of what you have to say.
- Make sure that everyone in the audience has one, and give them *time to read* it. Ask for questions, or uncertainties about content, before going on to talk about the hand-out. This is a good time to give the audience a chance to discuss what you have been saying, and to give you feedback.
- Always direct the audience's attention to a particular passage; then ask them to put the handout down and listen to you. Otherwise they will read on, and miss your discussion of the point.

- They will obey directions quite happily, but if you don't politely direct their attention it will fragment into small pools of distraction.
- Don't read while the audience follows – their eyes can read faster than your voice. They will get out of synchronization, and soon be bored and lost.
- Always control your audience's reference to paper. Take them into and out of particular points. Explain what a hand-out will be used for by saying something like, 'I shall refer you from time to time to examples in the paper before you . . . for the moment please put it to one side', or 'I want to make some general points first before we look at the paper I have just given you'.
- Never hand out full notes before you talk. The audience simply will not listen to you; they will read through the notes instead.
- If you tell the audience in advance that they will get prepared notes at the end, no-one will write anything down during the talk. If you don't tell them, they will be angry to find that they needn't have bothered to make their own notes. The only solution is never to write their notes for them.
- Try handing out a skeleton of the talk, and inviting the audience to make their own notes in the spaces provided. You must ensure that you progress evenly down this sheet as you talk, otherwise they will panic if you are only a third of the way down their sheet when the alloted time is nearly over.

Flip-charts

Flip-charts can be used like blackboards, but have a better visual impact. They are also paper, which is familiar, rather than matt-black wood. The main advantage they have over the blackboard is that they can be prepared in advance, and don't take time to draw during the presentation. But they retain the great advantage of the blackboard; they are dynamic, and can be added to during the talk. It is especially useful to add something such as the tick and two dots, which becomes a smile when the point is made. Figure 11.1 shows an example.

- They are excellent for graphs, bar-charts, circuits, plans and diagrams of all kinds.
- The disadvantage of flip-charts is that they are clumsy to turn over, and difficult to refer back to. They also tend to be messy, and the pens are nearly always squeaky.

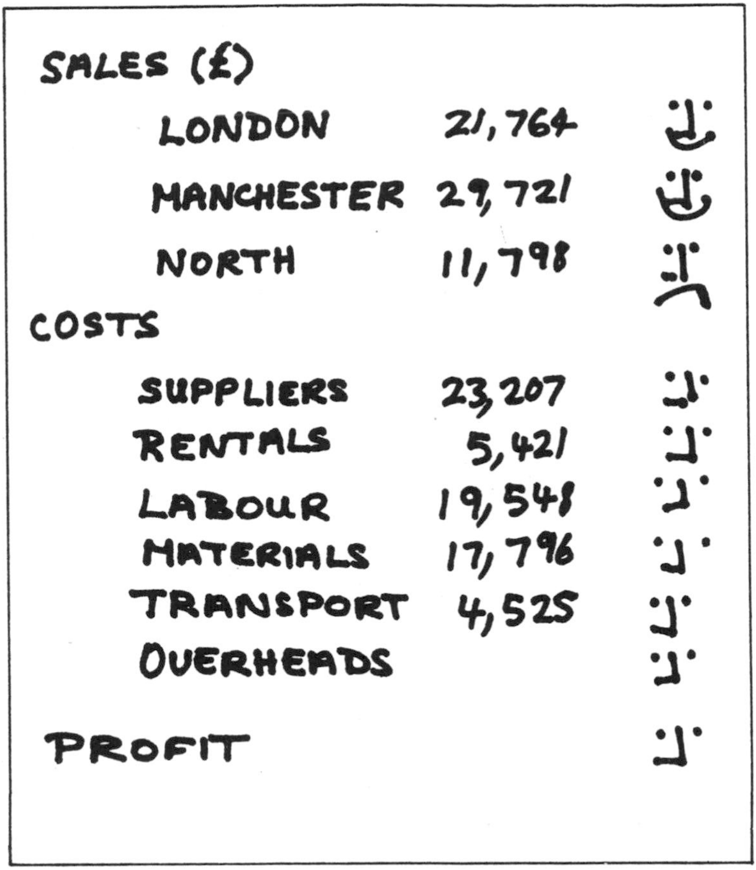

SALES (£)

LONDON	21,764	🙂
MANCHESTER	29,721	😄
NORTH	11,798	😐

COSTS

SUPPLIERS	23,207	🙂
RENTALS	5,421	🙂
LABOUR	19,548	🙂
MATERIALS	17,796	🙂
TRANSPORT	4,525	🙂
OVERHEADS		🙂
PROFIT		🙂

Fig. 11.1 A happy flip-chart

- They are usually too small for a big group. A seminar room of up to 15 or 20 is about the maximum audience they can be used for. For a group any larger than this, use an overhead projector.
- Check before you start that there is a good place to hang the sheets. If you often speak in a strange room, take two large bull-dog clips in your brief-case, so you can clip the charts onto a blackboard, or even a chair stood on the desk.
- Plan the drawing on a smaller sketch pad before you start the main one. Draw out the large drawing in light pencil before inking it in.
- Always use several colours, even if you go no further than red for the heading, and black for the list of key words.

- Don't use too many colours; three are quite enough.
- Never put too much on one chart; use several to build up an image. You can prepare them by tracing through from the complete drawing onto several sheets above it, one by one, to build up the diagram stage by stage as you flip over sheet after sheet during the talk itself.
- Check from the back of the room that the lines can be clearly seen, and the writing easily read.
- Use thick lines, if necessary by going over them several times.
- Never write or draw normal size. It may be visible to you, but it will be useless for the audience at the back.
- Always have the flip-chart covered with a blank sheet when you start talking. Have *two* blank sheets between each diagram, so that you have blank sheets after you turn over the diagram you have finished with. You can then introduce the next topic, without the audience being distracted by looking at new drawing. Two sheets are needed because most flip-chart paper is partially transparent, and the drawing underneath is tantalizingly half-visible to the audience.
- Ghost in lines and words you want to add during the talk in thin pencil. It will be quite visible to you, but invisible to the audience. You will then be able to draw, apparently free-hand, with great confidence.
- Flip-charts can be prepared by outlining the projected image of a slide or transparency.

The blackboard

- The blackboard is often the easiest aid to choose, because virtually every room you are likely to speak in will have one, but using it effectively requires more skill than it appears.
- The blackboard is good for spontaneous visuals, but many people are poor drawers.
- The great virtue of blackboards, and whiteboards, is that lines and figures can be altered on the spur of the moment, to remove mistakes, or to show the development of an idea or process.
- The audience can see the structure of the visual aid develop; it is dynamic.
- The board is excellent for noting key words, doing calculations, and emphasizing things to be fixed in memory. The audience can hear it, watch it being drawn while it is explained, and see the reminder until it is rubbed off.

- 'Chalk-talking' is a specialized art. It is better to be quiet, or briefly explain what you are doing, while you draw.
- It helps to work from a rough draft of the drawing in your notes.
- Try a dry run before the day. Solve problems of space, layout and perspective in advance.
- A good tip is to chalk in the drawing you need faintly when the room is empty before the talk. Then lightly rub it out. The faint lines will be visible to you standing right next to the board, but invisible to the audience. You can fill them in rapidly and surely, and mystify the audience by your skill in apparently free-hand drawing.
- You can prepare a complex drawing by projecting a slide in the empty room before the talk, and faintly tracing the outline onto the board.
- Using a roller-board, and positioning a drawing prepared in advance on the back, has the effect of covering up a visual aid until it is needed.
- Don't leave your drawings on the board after you have used them. Rubbing them off has the same effect as covering up any other visual aid. The audience can see the change of topic, and the pause while you clean the board gives variety. The act of moving round the board is a signal to the audience that they must clear their minds ready to receive a new topic; it reinforces the structure of the talk.
- Draw simply and clearly. Use simple language, and use key-words, never full, laboriously written, sentences.
- Don't talk to the board.
- Don't apologize for bad artistry; they can see it well enough!
- Don't over-crowd the board.
- Use colour if you can.
- Write clearly; it needs a lot of practice, or a faint ruling line, to write straight without getting a nervous droop to your lines of text.
- The board doesn't replace flip-charts. It can be used as well as these, indeed several different kinds of visual aid in the same talk increase the variety, and therefore impact, of the presentation.
- Try using two boards at once for different levels of points.
- Don't break up the talk by a futile hunt for chalk and rubber. Left alone in the room you would find them in seconds; in front of all those watching eyes you will look past them over and over again, to the audience's great delight.

- Make sure the board is clean before you start. Nothing is worse than getting to a point when you suddenly, and quickly, need a rough diagram, and finding your talk interrupted by clouds of chalk and sweat before you can go on. Get into the habit of always checking the board before you start to talk.

- It is good manners to clean the board before you leave the platform, so that the next speaker doesn't have this problem. It also gives an excellent impression of thoughtfulness, and good organization.

Projection

- All mechanical aids can be lumped together, because the same problems apply to all of them. They include slides, over-head projectors, epidiascopes and films.

- The disadvantages are real, and you must be sure that the advantages are worth the risks before committing yourself to 'electrickery'.

- Fuses and bulbs can blow, the slides can be in the wrong order or upside-down, the slide projector can jam, films can burn, and many overheads are too dim. You must, of course, always have a spare bulb and spare fuse with you.

- The room needs to be darkened. Is this possible, and what will the audience do in the dark? Go to sleep? Further, it is impossible to take notes in the dark. Nor can there be any eye contact with the audience in the dark; all non-verbal communication is blacked out. It is therefore difficult to remain sensitive to audience feedback. Also, the room can get stuffy. Ventilation and blackout are inimical. Darkness brings many problems with it. The solution, however, is not turning the lights on and off; it dazzles the audience when the lights are switched back on, and blinds them when they go off again. The solution is to go for a smaller, brighter image, and keep the room in half-light.

- All mechanical aids are bulky to transport, need technical help, are risky to operate, and can easily ruin a presentation if they break down. They can be very distracting and disrupting, and not surprisingly, most experienced speakers prefer to rely on appearance, gesture, and the repertoire of non-verbal signals they carry around in their own person.

- The advantage of projection, which we must remember since it often compensates for the risks and difficulties, is that it gives a large, brilliant, colourful image which has great impact. In drab buildings or drab surroundings, a colourful visual aid has a much greater impact. While in an interior decorated in modern fashion with bright primary colours, a black and white image will be more impressive.

Slides

- Photographic images have great clarity and impact. They are the next best thing to taking the audience there itself.
- But they are inflexible; they cannot be altered or changed.
- It is difficult to direct attention to part of a slide; the whole image is so brilliant and compulsive, that people are distracted by everything else in the picture.
- Drawings of machinery are better than pictures. They are also sharper and clearer, and distracting detail can be omitted.
- It is fairly difficult, and certainly clumsy, to refer back to an earlier picture in a slide show. You have to flash past all the earlier slides as well. It is better to repeat slides; have a duplicate where you need to refer to an earlier picture.
- Put in blank slides, or pieces of 2″ × 2″ cardboard, to provide the effect of covering the aid, or to signal changes of topic.
- Remember to turn the projector off before summing up and ending the talk.
- Always run through the slides just before the audience come in; check that the slides are in the correct order, and the right way round.
- Try drawing on slides with a felt pen. Circles, arrows and under-linings can be very effective when projected.
- Use a pointer, not fingers, or the slide is obscured, and the projected image will appear comically distorted on your forehead.
- Have slides in a meaningful order, starting with a distant general view, or a map, and moving inwards towards the detail. This gives a sense of context and scale. Also make the movement around the object logical, so it symbolizes a possible tour. It prevents the audience getting disorientated.
- Don't use too many slides. A lot of rapidly changing slides are distracting. Some people get motion sickness, and most will feel disorientated. A large amount of invisible mental process-ing has to be done before a new image is orientated and decoded in the viewer's mind. If these huge demands on processing time

are repeated too often, the viewer gets tired and his perceptions fail. Having images which represent moderate, and logical, changes of view-point from the previous ones can help. But careful limitation of the amount of material offered is essential. Five slides are quite enough for a ten minute talk, and a dozen will fill half-an-hour. Be ruthless in rejecting all but the most essential images.

• Check focus, screen position and visibility by trial projection before you start. Do not only check a horizontal slide – an essential bit of a vertical picture may be missing, or projected grossly out of scale on the wall above the screen, in the actual talk. All these little errors contribute to the audience's dismay at an ill prepared presentation.

• Put the screen to one side, so that it shares the centre of attention with the speaker, as shown in Fig. 11.2.

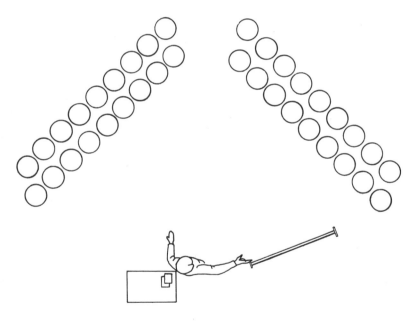

Fig. 11.2 Placing of screen and speaker

• You must have either remote control, or a reliable projectionist.
• Auto-focusing projectors are good, and a must for speakers who use slides often. They remove one more problem. I once used one

which went wrong, so the focusing had to be done manually. For the short period before it was repaired, I had continual hostile criticisms of the content of the slides from audiences. When the focus was reliable again. I had no comment about it, but the hostile criticisms of other things ceased. Audiences are usually naive about the true sources of their dissatisfactions.

● Many of the points about flip-charts also apply to slides.

● Make sure there are no distracting objects like a person's leg, or an unusual motor car, visible at the edges of the picture. Black masking tape can be used to cut down the image size. Often a slightly masked image does not seem any smaller than the others. Severe masking can create an impressive effect. A small, bright image of just the object being discussed is exciting.

The overhead projector

● Many of the same points apply to overhead projectors as were made about slide-projection.

● The greatest advantage is that it is possible to draw on the foils while they are projected, so the presentation can be dynamic, as well as enlarged by projection.

● Leave arrows to show direction flow, legends, and connections to be added during the talk. It makes the point much clearer than a complex, static picture. Flows in chemical process plants, feedback loops in electronic circuits, outstandingly good (or bad) figures in a table of results, can all be effectively added at the point when they are discussed. Leaving them out raises curiosity in the watchers' minds, and they are remembered much more vividly when they are drawn in.

● A complex image can be built up from several layers of transparent foils, selotaped together along the top edge, and folded over one on top of the other.

● Overlaid transparencies can be made to slide, or rotate, to show movement.

● Always use cardboard mounts when rolls are not used. It makes the flimsy and sticky foils easier to handle. With nervous fingers, it can be surprisingly difficult to separate a pile of foils, and grasp the one you want.

● Check that the projection is bright enough with the planned lighting before you start.

● A disadvantage is that the light from the platten is in the speaker's eyes. Stand well away from the machine while talking and take a pace forward to change slides.

- Point to the platten, not the screen; you can then continue to face, and look at, the audience.

- Use a pencil or pen to point with. Leave it stationary on the platten while you elaborate the point it is indicating. Do not hold it while you talk, because a tiny tremor will be magnified into apparent terror on the screen. Step back, leaving the pen in place. Then step forward, re-position it, and step back again to continue to have eye-contact with your listeners.

- Use the switch provided on the front of most projectors to switch it off as often as on, and certainly switch it off while changing foils, and between each point. The machine should always be switched *off* by the time you finish talking. If you find yourself closing with the light still on, put a large reminder, in red in your notes, to switch it off before the summing up in the next talk you give.

The opaque projector

- Sometimes called the 'epidiascope'; because the opaque projector is usually not as bright and clear as other forms of projection, it has fallen into disuse. But it does have the single advantage that it can be used to show printed material and books.

- It is better to photocopy the page onto a transparency. Most agencies can do this.

- Only useful for *ad-hoc* presentations. Anything which can be prepared in advance, or needs to be used more than once, should be turned into a transparency or a slide.

Film

- Cine film is a complete medium in itself. The talker has nothing to do while a film is being shown, and can no longer interact with the audience.

- A film replaces a spoken presentation, it doesn't aid it.

- The only function of the speaker is to introduce the film. There is no doubt that the film medium, if well made, can give a very powerful impact to the presentation of information. But they miss the personal element provided by a speaker who is expert in the topic. You can't ask a film questions.

- Don't try to talk *after* a film. It is such a powerful medium that

- you will be seem very flat in comparison. Film is an impossible act to follow.
  Many speakers use other people's material, which is often inappropriate to the point they wish to make.
- Material filmed for even a slightly different purpose can be confusing and disappointing for an audience.

Audio aids

- Many speakers forget that their aids do not have to be only visual. Record players, radios, casette- and tape-recorders can also be used, and their novelty or appropriateness may have real impact.
- Always edit a tape, and extract just the message you want onto a separate tape or cassette. This avoids losing your place, or being fooled by unreliable tape-counters.
- Check that the sound is clear, and loud enough to be heard from the back of the room.
- If used in a large or echoing room, those sitting at the back may find the sound impossible to disentangle from the echoes.
- Use coloured leaders on reel-to-reel tape to indicate gaps between examples. Otherwise, use a long blank stretch to ensure easy location.
- Short, illustrative noises can too easily sound like the Goons.
- Be careful, when using an unusual aid such as sound, that the illustrations you offer the audience are relevant and not distracting.
- Audio aids can be used to create the effect of a team presentation when there is no other person present. Recorded points can introduce a variety of voices, and may help the talk maintain interest.

Models

- Models, samples, or even better the real thing if it is portable, can be very effective visual aids.
- Models need not be to any particular scale, and are better as a demonstration aid if they are not very detailed. The lovingly-made work of craftsmanship may be a delight, but will also be a distraction.
- Simple, large, chunky representation of the basic geometrical and topographical relationships are best. Mechanisms should be clear and simplified. If they are moving, they will help under standing.

- Good examples are the architectural models which omit doors and windows, and represent buildings simply as coloured blocks of wood.

- Samples are fine, but do not hand them out, unless you have one for everybody, and a real point to make. For example, in one talk I saw, samples of silicon rubber were given to the audience during the talk. It is a fascinating material in the raw, and the small samples were bounced, smelt, sat-on, chewed, rolled, thrown and enjoyed by everyone. No-one listened to the rest of the talk.

- The model must be worth the effort. One speaker produced a large, floppy but painstakingly drawn model of a slide-rule, when several real ones would have worked better.

- If you use the real thing make sure it is not too large, or too small. The person at the back must also be able to see it.

- Make sure it comes apart, or works, before you use it. I well remember the speaker who produced a full kit of diver's breathing apparatus to illustrate his talk on underwater sport. To make the point that the apparatus was safe to use, he chose to show how easy it was to dismantle and clean. Picking up the apparatus, he tried to demonstrate. But no matter how he fought, it refused to come undone. The audience was holding its breath!

- Keep it covered up until you need it; under the table, in a bag, or under a sheet. For example, one speaker, when talking about railways, managed, with the help of several friends, to bring along a model steam engine nearly two meters long. During his talk, the engine sat on the desk in front of him. It was admittedly a beautiful model, and everyone spent the time looking longingly at it instead of listening to the talk. I still can't remember anything he said.

- Put the aid away afterwards, before summing up. It is useful to promise the audience another look at the end of the talk, though you must not then keep them waiting too long.

- Don't hide it behind your hands during the demonstration. Hold it well up. Make sure it is the right way round for them to see. You will have to lean round it to see what you are pointing at. Figure 11.3 shows the correct position.

- The real thing has a surprising psychological impact. One general interest talk on artificial hip joints, aroused undivided attention when, half way through, the speaker produced a selection of real nylon and chrome joints.

Fig. 11.3 Speaker holding aid correctly

Summing up the advice

I have offered a range of detailed advice about individual aids in the last few pages. But whatever aid you use, there will be new ways of using it which are particularly appropriate to the subject of your talk, and which therefore cannot be predicted in a book of general advice. The art of visual aids is not a rigid routine, but a watchful and thoughtful sense of relevance. In particular, other people's talks will give you ideas. They will also give you warnings. The best general advice is to keep a careful watch on the principles outlined at the beginning of this chapter. Ask yourself whether the visual aid you propose to use does the job, and what the real purpose of using it is. If you cannot justify the use of the aid, reject the idea as a distraction both to you and to the audience.

Prepare with equal care the aids themselves, the notes to support them, the situation and the room. The key to success in all types of speaking is preparation; Baden Powell's motto, 'Be prepared', is as relevant to speaking as it is to war. The basic point is doubly important when using visual aids, for whereas you can cover up cracks in your verbal presentation by a few *ad hoc* remarks, the wrong pictures, upside down slides, and illegible flip-charts always stand out.

One of the last pieces of advice is also one which will help to calm your last minute nerves. Always do a last minute double-check of the aids, even if you know everything is fine. This check should show nothing wrong. If it does uncover problems the earlier checks are not good enough, and you will have to improve your preparation technique. If it doesn't uncover problems, as hopefully it won't, it will boost your confidence, and thereby calm your nerves. The use of visual aids will then have contributed to a more effective talk, both because they have helped the speaker, and because they will help the audience to stay alert, interested, and attentive.

Notes to chapter eleven

1. Ruth Beard, *Teaching and Learning in Higher Education* (Penguin, 1976), p.115.
2. *Lotus 1-2-3* is a trademark of the Lotus Development Corporation, USA.
3. John P. de Cecco, *The Psychology of Language, Thought, and Instruction* (Holt, Rinehart and Winston, 1969), p.373.
4. M.L.J. Abercrombie, *The Anatomy of Judgement* (Penguin, 1979), p.38.
5. Donald Bligh, *What's the Use of Lectures?* (Penguin, 1971), p.114.
6. M.D. Vernon, *The Psychology of Perception* (Penguin, 1982), p.54.

SUMMARY SHEET

Chapter eleven – Visual aids

Visual aids help the audience listen because they:
— get attention
— give variety
— make the talk easier to remember
— save time when explaining complicated points
— show things which can only be explained graphically
— give the speaker something to do
— remind the speaker of what comes next.

The disadvantages are that they:
— interfere with personal contact between speaker and audience
— can't tell the whole story
— can't show abstract concepts, such as entropy
— distract the audience if not chosen carefully.

When preparing visual aids check that they are:
— appropriate to the point you are making
— perfectly clear, and not confusing
— simple and bold
— big enough to be seen from the back
— not distracting, or with irrelevant details
— covered up between each topic
— Seen long enough to be read and absorbed
— not talked to, and recorded in your notes

12

Persuasive advocacy

All talks are persuasive

Most talks are persuasive in some way; few are just for information. In every talk the speaker must at the least persuade the audience to listen, to see his or her point of view, and try to understand the ideas and information he or she is offering them. In a persuasive talk, though, people must be motivated to get things done, to act, or to spend money. In other words, the speaker must ask for something. Getting something done, causing actions, requires movement; and to create movement, momentum must be generated. Persuasive speaking is the art of generating action, not just knowledge, in others.

Perhaps the first point to make about persuasive speaking is that skill as a persuasive speaker is not something nefarious. Persuasion is a familiar, regular, and important human activity: 'in the largest percentage of all human interactions the basic decision-making tool is not fighting, not biting, not roaring, not hissing, but persuading.'[1] It is strange, then, that offering to teach this very basic and universal skill suggests to some people a calculating, rather underhand, attitude to human nature.

Of course, persuasion has a bad reputation as a skill because of its association with propaganda; but persuasion is not necessarily devious manipulation. Persuasion ought to be harmless; its job is to give other people an opportunity to understand, and if necessary resist, what is being proposed. As with any human tool or skill, persuasion can be misused. But responsibility for the misuse cannot be laid at the door of the skill itself. To eschew the art of persuasion in a good cause, leaves your audience defenseless against the skills of persuasion used in a bad cause. I think, also, that our society has become used to persuasion, and adept at perceiving it for what it is.

Our economy is based largely on advertising, and the skills of marketing. It would be a very straight-laced person who thought all marketing was a crime.

Some advertising, no doubt, is less than honest. But most is responsible, and often entertaining. I, for one, am prepared to believe the professional marketing man when he points out that only good is done by presenting the best case for something. And this is what, in the end, persuasive speaking is doing. One cannot, short of violence, make anyone do anything completely against their will. Nor, in the end, can anyone less than a magician, persuade a person to do something they would rather not do. But a good speaker can make the best case possible for what he or she has chosen to support, and can help unite a group of people for fruitful action. Persuasion is not supernatural: it is a useful set of skills for the speaker.

Good persuasive speaking is not easy, which is why I have left it until this late in the book. Persuasive talks require all the skills used in informative talks, but with an advanced level of diplomacy, and strategy. The basic tactics of speaking are the same as before, but there are also new strategies to learn, if the audience is going to be moved to action. Their natural inclination is to cling to the status-quo, but for the persuader status-quo-itis is a disease which must be contested. He or she cannot bore the audience into submission. The speaker has to contend with the audience's limited time and attention with great skill. If he or she mis-plans the timing, or numbs them into inattention, they will fail to achieve anything. Everything is lost unless the message is complete in time, and the audience have been persuaded to listen. Their agreement comes only after these things have been achieved.

So all the skills of speaking I have discussed so far must be practised if you are to persuade. There are other skills, too, which you must acquire if you are to impart motivation, as well as information. You must be aware, for instance, that the audience's attitudes towards you are a large part of their willingness to be swayed. The speaker must also contend with the risk of the audience feeling, in the common phrase, that: 'what you are speaks so loudly we can't hear what you say'. The speaker must be credible enough not only to win their attention, but also to be able to persuade them to act. Because of these additional demands on his speaking skills, the persuader will find that good tactics are especially vital.

Tactics

I divide the tactics of persuasion into two main groups. Most people learn while they are still school-children that there are two broadly different ways of trying to get what they want. They either ask for it, and then explain why, or they give a whole lot of reasons, leading up to asking for it. In adult life, too, there are two basic tactics; either say what you are there to get, explain why, and then repeat the request; or describe the problems first, and only ask them to agree to the proposal at the end. These two techniques have widely different psychological effects on the audience. Broadly, one should *always* say what one wants first, because otherwise the audience may misunderstand the implication of the points you make. If you say that the new chemical process is expensive, but has a high purity output, the audience may understand you to have said that the cost was too great, and expect you to ask them to approve the cheaper process. You, on the other hand, may have meant that the purity compensated for the cost. Such misunderstandings are surprisingly common. People cannot guess what a speaker means; they need to be told. If you had started with the clear statement that you were about to make a case to justify the more expensive process, then the evidence would have been understood in the right light. People cannot evaluate evidence if they do not know what it is meant to prove.

The normal order of announcement is therefore best for most persuasive talks. In this, persuasion does not differ from informative talks. But there are two important differences in tactics. In some cases it may be better to delay explaining what you are asking for until the end of the talk. Which are these cases? Simply, where your request is likely to arouse instant, unthinking, and blind opposition. If you make the request first, this opposition may prevent the audience from listening to the reasons you give. Here it is better to survey the arguments, balance the pro's and con's, and only then work towards the conclusion which by then should be both obvious, and well supported by evidence.

You may have noticed both these tactics being used. The most familiar example is the television charity appeal. I recommend watching these from time to time, with an eye on the techniques used, for as in most speaking you can learn a lot by observing others, and observing the effect of their tactics. A rough survey suggests that television appeals are divided half and half between those which make a direct appeal, and then describe the need, and those which describe the needs, and work up to an appeal for help. Watch a few appeals, and think about the different effect of these techniques.

Say what you want

The second way in which a persuasive talk is different from an informative talk is the need to be more than usually clear about what you want. I have observed many hundreds of persuasive talks, some as exercises, and some as real life presentations. My technique when teaching persuasive speaking is to get the audience (usually a group of ten or so other learners) to vote on the proposition immediately after the end of the talk. Of those talks where the proposition was voted against, more than half had failed because the speaker had not made what he wanted clear. If asked: 'why did you vote against the proposition', the group typically replied: 'we didn't know what he wanted'. Failure to say what is wanted sounds bizarre, but there is a simple cause. The speaker knows what he or she wants, has lived with the idea for some time, and is perfectly familiar with the main point. They therefore fail to realize that not everyone is as familiar with it. Feeling that they shouldn't labour the point, and that it is fairly obvious anyway, speakers restrict what they say to the points which support the issue. But they never make the issue itself clear. Probably the stupidest reason for failing in a persuasive talk is that the audience didn't understand what you want. Remember that they are not familiar with the point: to them it is new, whereas to you it is already accepted. Make a special point of giving one clear, simple, unmistakable statement of what you want. The best time to do this is at the end of the talk, because then it will have most impact. There is a psychological advantage in making a clear request the last thing in a talk. People find it easier to agree, than to disagree. You will often have found yourself nodding agreement with political gossip which is quite different from your own views because you don't want to make the effort to disagree. The same mechanism works in an audience. If they are asked, directly, positively, and unmistakably for something, it will require an effort to deny it. Use this simple inertia to help achieve your point. End your talk with a direct request to which they must respond, either by agreement, or positive disagreement. Most of the successful persuasive talks I have watched have succeeded for just this reason. At the end of the talk all the reasons, plausibly and persuasively presented, are fresh in the audience's mind. A powerful, direct request at the end has then created a momentum in the audience's thought which has carried the point.

When you give a persuasive talk, then, the basic principles of clarity become even more important. So do the basic skills of creating interest, and holding the audience's attention. You aim to change people's minds, get them to accept new ideas, take action, and spend money. This can't be done by boring them. Only by interest and involvement can the momentum be generated to achieve action. Everything I have said in earlier chapters about understanding the audience, relating to them, creating variety in voice, gesture, and manner, and timing the presentation are now even more important. A persuasive talk typically fails for three different reasons, and it is difficult to decide which is more important:

1. failing to make clear what you want;
2. boring the audience, and losing their sympathy;
3. having a poor case to present.

I have deliberately put the 'poor case' last of the three. People do not usually support cases which have no advantages, or which are in no sense plausible. My experience is that a poor case is the least common reason for a talk failing to persuade. It is lack of technique, failures of skill, which are the common reasons for failure.

There is some interesting research evidence which supports this advice about tactics. Research shows that overt behaviour is affected most if specific actions are recommended. Earlier arguments have most effect, so the most appealing part of the argument should come first. It is useful to start with statements with which the audience will agree, in order to win its confidence. But it hardly needs psychological research to support these obvious points. Simply watching the way other speakers operate, and noticing what works and what doesn't work in your own talks, will provide enough evidence to convince you of the importance of making what you want abundantly clear.

Credibility

Essential evidence for the decision on tactics is your knowledge of the audience. As with any talk, it is the audience who decide whether you fail or succeed, and most attention must be paid to them, their needs, interests, and attitudes when preparing the talk. You must ask yourself whether they respect you, whether they are friendly, or unfriendly. Try to judge, also, what preconceptions they are likely to have about you, and the case you are going to present. Hovland and

Janis, in their classic books on persuasion, suggest that there are seven characteristics of 'persuadability' in the receivers of a message. These are: low self-esteem, perceptual dependence, social isolation, richness of fantasy and sex. Two remaining factors militate against persuadability: other directedness, and authoritarianism. I am not suggesting that you should put your audience on the couch, and psychoanalyse them. But you should think about their likely reactions. If they are known to you, it may be possible to estimate how many have easily persuasible personalities, and how many are authoritarian and rigid.

Analyse firstly their confidence in you, and their respect for you as a person. Be careful never to patronize, or lose their respect by not respecting them. The second area for analysis is the extent of their current knowledge? Do they know about the subject, or is it new; are they laymen or experts? You must finally ask, can they act? and plan your tactics around those who can. It is also necessary to ask, how does the change or action you are proposing affect them and their status? Remember the importance of group loyalty. Individuals are afraid to move against a group from within it. Usually no one individual wants to take responsibility. An effective speaker must make an effort to fit in with the group behaviour norms, loyalties, dress and style.

The next questions must be asked about yourself. Are you sincere and thorough? How much self-involvement and self-interest do you have in the outcome? I discussed non-verbal signals at length in chapter nine, but they are especially important in persuasive speaking. They are powerful and involuntary, and almost impossible to fake. Unless you are aware of the unintentional non-verbal, message you are giving to your audience you may well undermine the persuasion. This is a point I will return to later in the chapter when talking about advocacy.

This whole subject has received a lot of attention from experimental psychology, under the name of credibility research. Thus, credibility research suggests that the persuasiveness of a lecture varies with the credibility of the lecturer. The effective credibility of a speaker depends on his or her expertise, trustworthiness, fairness and intentions as perceived by the listeners. Of these factors, research found that fairness was more important than expertise and intent. All the research on credibility shows that features of a personality that are clearly irrelevant to the subject, such as appearance, prestige, ability at sport, and membership of a prestigious minority group, affect a speaker's persuasiveness.

These are results which most people would have predicted; anything which upsets your credibility will reduce your ability to persuade. The most important factor in credibility is that you should be perceived as honest and fair. Openness (which is supported by clarity), and friendliness (shown by smiling and looking at everyone) will have as much impact as your knowledge of the subject.[2] Another psychologist points out that: 'Face-to-face communications are multivariate, and when we are being sincere, proxemic (spatial), postural, gestural, extralinguistic, paralinguistic and linguistic cues will be in harmony. When we . . . are putting on a performance, we have to control and monitor all cues presented.'[3] Whereas it is, of course, possible to fake these signals (indeed, actors do it as a career) it is surprisingly difficult to get them all right.

We are very aware of the moods and inner thoughts of people we are talking to. If someone is lying, it is often obvious. Even if you cannot pin down exactly why, there is an unease, a suspicion, lurking in the mind. This suspicion is caused by non-verbal signals which do not tie in with the claimed message, although people are quite unconscious of their perception of these non-verbal signals. So unless you are a consummate and practised actor, the best advice when speaking persuasively is that it is wiser to be sincere. It is best to persuade people about what you believe in: it is usually obvious when you are advocating something you are not yourself convinced about.

Be full of tactics

When you have considered the audience, and your own motives, there is a third factor which must be taken into account in planning tactics. Think about the organization to which the audience belong. Management analysts discover that some organizations depend on 'conflict resolution' as a generalized method of solving problems. This means that they make decisions with the aim of minimizing the amount of conflict in the organization. Decisions are a sort of averaging of the views of all the members of the organization. It is as if the principle of the 'least unhappiness' was being applied; no-one gets exactly what they want, and no one is totally ignored. The technique to apply in such an organization is to stress the claims of those members of the organization who support your case, and try to show ways in which damage to the claims of those opposed to it can be minimized. In other words, your tactic is to present your case as a compromise which will increase harmony, rather than increasing conflict.

In most organizations it is wise to tailor your persuasive message so that it has something for everyone. Persuasion implies that something needs changing, and change will induce conflict. As many people as possible must be persuaded that the change will bring compensating benefits. Such points are, as with most of my advice on speaking, fairly obvious. If they were always taken into account, it would not be necessary to mention them. But my experience has been that all too often the obvious is overlooked, the simple and sensible is ignored, and the result is failure. It is therefore wise to make a deliberate and explicit assessment of the organization when planning your persuasive talk.

As I said at the beginning of this book, nothing clarifies thought as effectively as writing it down. Why not make a simple analysis on a sheet of paper, under three headings: audience, speaker, organization? Hopefully, you will be able to jot down the obvious points with no difficulty. If your talk is to succeed, it should be a trivial task to make this analysis. I am prepared to wager, though, that most of the time the actual task of making these analyses explicit will be far less trivial than you expected. Usually, one finds that there is not enough information to make decisions. More thought, and more information, must then be added.

The simplest way of finding more information is to talk to the people who organized the talk. If you are talking in your own organization, ask the manager or colleague who arranged the talk who he expects to be in the audience. There is no need to make a great drama out of this inquiry, but you should be able to inquire about these people's attitudes to the type of case you are going to make. If you are talking in a different organization, contact the person who invited you, and make inquiries. Whenever I am invited to give a talk, I telephone a few days before, to make a general check. This usually takes the form of first confirming the date and time, secondly asking for any equipment such as overhead projectors, and thirdly inquiring who they expect to be there. It is perfectly easy to explain that it will help you to know how big an audience to expect, and roughly what composition. Will they be mainly managers, mainly people working on a similar topic, or mainly people not directly concerned? Do you know of any strong opposition to this type of scheme? And so on.

The situation in an organization or group prior to beginning the persuasion will greatly affect its outcome. What decisions, debates, changes and rejection of new ideas have occurred recently? Careful analysis of the present balance of interests, and a thorough knowledge of past movements in ideas and actions is vital. It would be useless to try to persuade an organization to adopt a policy if a similar one had recently been rejected. But you may also draw benefit from a previous proposal. On the one hand, the differences between the new proposal and the last one may help it succeed. On the other hand, the similarities will make the supporters of the last proposal support the new one. And you may win over some of those who nearly supported the last proposal. They may finally be drawn over to the side of change. The virtues of the last proposal may be added to the virtues of the new proposal. None of these tactical advantages are available to the speaker who has failed to consider the history of the situation.

The motivating forces

It is often said that people are driven largely by simple motives. I think this is mainly true, and a knowledge of these motives wlll make it easier to select the right ways to motivate people. If you can attach your proposal to a strong and familiar motive, it will be drawn on to success with little effort. If, on the other hand, you make your proposal appear to be in opposition to people's natural motives, it will be swimming against the tide, and will require much more work to make it succeed. What are these natural, common, motives? It is optimistic, but naive, to believe that simple logic is the sole, or main, influence on decisions. Experience shows that people are motivated by habit, emotion and intellect in that order.

Habit

Habit is one of life's most useful mechanisms. Most of our actions, after all, are repetitions of the same thing. Habit is a useful way of saving time and energy by solving repetitive problems with relatively little effort. Do not disdain habit; the art is to use it. Most people resist change because it means they must change their habits of work, and habits of thinking. Often this resistance can be dealt with by showing the similarities between the new proposal, and the habitual ways of working or thinking. One clever technique is to persuade people that the real habit is the new one. Persuade them that innovation has been a habit, that excellence is a habit, that adventurousness is a habit for this particular group or organization. If we can see that habit is only a way of satisfying our needs we can use it on our side. The persuasive communicator increases his chances of success if he can establish that a proposed change is only a small extension of an established goal or no actual change at all. Most organizations dislike uncertainties, and try to reduce them. The uncertainty model of persuasion claims that speakers are more successful if they explain their new suggestions as part of the organization's tradition.

We often overrate fresh thinking, and assume habit is something to be ashamed of. It is one of the curious by-products of the Romantic revolution that we value habit and convention so little, and independence and fresh thought so much. But Alfred North Whitehead insists that thought is a rarely used tool:

> It is a profoundly erroneous truism, repeated by all the copy-books and by eminent people when they are making speeches, that we should cultivate the habit of thinking about what we are doing. The precise opposite is the case. Civilization advances by extending the number of important operations which we can perform without thinking about them. Operations of thought are like cavalry charges in a battle – they are strictly limited in number, they require fresh horses, and they must only be made at decisive moments.[4]

Habit, then, is a most valuable tool. Rather than despising it, and emphasizing what is new in our proposal, we should use habit as an asset. Habit is one of the main levers in persuading an audience.

Emotion

The second lever which a speaker can use to help persuade the audience is the power of emotion. All emotions come from personal needs and fears: they are the biological motive power. Because we live in a complicated society, many of our physical emotions have been translated into intellectual and mental driving forces and ideals. Most people rarely meet raw fear, except when watching a horror film. So emotion is a wide, but shallow, force for most civilized people. What motive forces can be included under the heading of emotion? A common summary of the so called ladder of needs is:

Body (I can't breathe)
Safety (I must protect myself)
Social (I want to belong)
Ego (I'm terrific)
Development (Better than last year)

Few persuasive talks are going to appeal to the most basic needs – they would fall into the category of threat and torture rather than persuasion if they did. The exception is the sex drive; which is lavishly over-used by advertisers. The reason is obvious; it is the only one of the basic bodily emotions which is still present in great strength in civilised people. I'm never sure what attitude to take to the use of mild sexual titillation in advertising. My (many) feminist friends argue that it is degrading, several of my (mainly male) friends enjoy it, and I suspect that I really think that it matters little. But I *am* sure that it is out of place in any serious persuasive talk. To try to adopt the techniques of advertising usually leads to failure, for it requires great skill and a polished advertising technique. Stick to the motivating emotions higher on the ladder of needs.

There are several general points which need to be made about the use of emotional appeals in persuasive speaking. The first is that emotion is more dominant in groups: more hidden in individuals. It is this factor which enabled demagogues, like Hitler, to motivate a whole nation to criminal acts. As a lesser demagogue, you may well find it easy to stir stronger emotion when talking to a group than when dealing with an individual. But emotions are always unpredictable, and it is safer not to try to stir strong, explicit emotion, until you are experienced and know your own power. Even political leaders, these days, rarely try to stir strong emotion.

One reason why emotions are stronger in groups is often said to be because each individual feels his personal responsibility diminished by the group's as a whole. Another reason is undoubtedly the fact that emotions are contagious. The exact physical mechanism for this is still a subject of interesting research. Candidates for the transmission medium include pheromones, the complex scents which some animals use to communicate. There are even claims for telepathy, though they never survive experimental testing. But whatever the mechanism, there can be no doubt of the fact. Even yawning (not a very strong emotion) is contagious; the powerful emotions like fear (panic in a crowd), ego (patriotism in a crowd), and anger (rioting in a crowd) run like wildfire in a group.

There are three ways in which this emotional force can be used. The first is contrary suggestion, the dislike of the competing proposal, which is much stronger in a crowd. It is possible to raise derision, disgust, and boredom more easily in a group than in an individual and this, as we shall see, can be used to your advantage in persuading a group to do the opposite.

Secondly, you need to be aware that groups hate a strong contrary emotion in an individual, if it is discordant with the mood of the rest of the group. As long as you carry the group with you, it is all well and good to express strong aversion to the opposite case from your own. But if you sense that the group is not sympathetic, you should quickly revert to neutral matters. Exactly the same applies to positive emotions of excitement, enthusiasm, and pride. Nothing seems more ludicrous than a speaker wallowing in enthusiasm, which no one in the audience shares. Listeners feel suddenly strangely detached, and remote from an emotion they do not share. A whole group experiencing a different emotion from the speaker will offer no sympathy at all. As always, the speaker's sensitivity to the audience is the key to success. In persuasive speaking, especially, this sensitivity must be great. Unless you have a practised sense of the audience's reactions, leave strong emotional appeals for others.

The third way in which emotion can be used by a persuasive speaker, is by using the more homely emotions of sympathy, and interest in oneself and one's neighbours. Everyone is interested in the personal, and in every kind of talk, personal anecdotes help. A rule-of-thumb is that the closer the proposal can be made to the audience's own lives the more real it is, the more remote it is the more indifferent they are.

Intellect

Many people like to think of themselves as rational creatures who, even if they do occasionally give way to emotion, behave and decide on largely rational grounds. Sadly, there is little evidence that this is the case. Plato thought of the emotions and the intellect as two horses, one white and one black, drawing the same chariot. But Nietzsche in nineteenth century Germany realized that the intellect was little more than the emotions' servant. He compared the emotions to a very powerful, but blind, man. The intellect was weak, but could see; it rode on the blind giant's shoulders, and pointed out the way. But the intellect could not decide where the emotions wanted to go, it could only tell the emotions how to get there.

Most people *think* they are being guided only by reason. But reason is clever, and it can always find a reason for what the emotions make it do. Psychologists call this 'rationalization', and it is one of the human being's most powerfully developed faculties. But the picture is not all black. D.H. Lawrence, among many other writers, taught the wisdom of the emotions, and the wilful errors of the puny intellect.

This chapter is not about to turn into a sermon on modern literature and thought. But I do think it is worth taking time out for two paragraphs to underline the real relation between the emotions and the intellect. Where imperious bodily emotions are concerned, the intellect has precious little persuasive force, unless it can show a better way to satisfy those emotions. Even where the milder emotions of pride and reputation are concerned, an intellectual argument which denies all these feelings is unlikely to succeed.

The moral is, intellectual and rational arguments work best when they work in harmony with the emotional arguments. If you can show that your pet scheme is not only reasonable, but is the best way to satisfy the majority of the higher emotional goals, then you are likely to succeed. We may like to think that intellect guides us, but most people are swayed by emotions, especially when they have rational support. The fact is that whereas we like to appear to argue rationally, we use reason mainly to give local colour and credibility, much in the way that politicians use statistics. When it comes down to it, reasons can be adduced for almost any position. The good persuader is someone who recognizes the power of emotions, but recognizes also that everyone wishes to appear rational. He or she satisfies both needs if he can appeal to the emotions using reason as the vehicle.

What I have said is not an argument for irrationalism. It is an argument for creative use of appealing reasons. When using rational arguments, recognize that only professional and intellectual audiences are likely to be able to judge the arguments on rational grounds alone. Experience shows that many people can not distinguish between a statement which merely describes what is there, and an inferencial statement which describes the consequences or deductions. For this reason, it is better to try to change the audience's perceptions, rather than disputing ideas and opinions on logical grounds. The speaker must aim to change the way the audience see things. Changing these 'frames of reference' requires an understanding of their strength, and the way they are formed:

> The fact that people do not discard the frames of reference they have acquired but adapt those frames as new information is available, makes sweeping persuasive change difficult to obtain. . . . If an individual believes that members of a particular minority group are dirty, lazy, stupid, and troublesome, meeting someone from that minority who is not dirty, lazy, stupid and troublesome will

not result in drastic changes in the original frame of reference. . . . the belief pattern might now be: 'Members of that minority group are dirty, lazy, stupid and troublesome, except for Joe Jones, who is almost like me'. Many encounters with many Joe Joneses are neccessary to produce basic changes. . . . new information that is contrary to that frame will produce few noticeable changes in behaviour. In fact, it may seem as if the material is not being received at all.[5]

The levers of persuasion

An understanding of the way people's motivations and perceptions work, then, is a great asset for a persuasive speaker. What, in details, are the tactics which you can derive from this general discussion? I find it helpful to recommend seven 'levers' which can be used to change the audience's mind on an issue. If you try to include an appeal to most of these levers in every persuasive speech you make, you will be surprised at their effectiveness.

1. Self-preservation

One of the strongest of human motives is the desire to protect one's job, group, safety, organization, family, and children. Just because it is so strong, it must be appealed to with care. But there is no doubt that if one can show that the proposed scheme will help to safeguard the profitability, and therefore future security, of the company and all the jobs in it, one is already half way there. Similarly, a scheme which will increase the security of local families and children will get a favourable hearing.

2. Possessions

The second strongest lever is the appeal to possessions. This is not just buying people's agreement by offering material benefits. It is the whole complex of feelings associated with wealth, profit, savings, and pride in spending power. Persuasion is increased in effectiveness if it can show that the individual or the organization will, as a result, be able to take increased pride in its physical and financial standing. By contrast with pride in owning a desirable object, saving is a negative virtue. The best way of persuading people to save, as many financial and savings institutions have discovered, is to suggest that what is being saved is the power to do something else in the future.

Advertising, as always, provides some of the nicer examples of deliberate (and often blatant) appeal to pride in possessions. Here is an example which was first pointed out by my colleegue, Peter Hunt, of an advertisement for a book club:

> Imagine the sheer pleasure of displaying them in your own home: an abiding complement to your good taste, an investment for the future, and a constant source of reading pleasure.[6]

3. Power

Nietzsche believed that the desire for power was the main human motivation. The need to be able to control the world, to provide safety for oneself, and increase opportunities for pleasure, is certainly a basic drive in all organisms. Human beings add to that a strange pleasure in dominating others. Everyone has a desire for influence over others, even if only through the effect of arousing their admiration. Perhaps one of the greatest powers is held by people who are so respected that the others try to emulate them. Pop idols have this kind of power, as do royalty, and famous actors and actresses. One effective persuasive technique, then, is to convince the audience that the scheme you are proposing will increase the amount of influence they will have on others. The assertion that if the new chemical process is adopted, it will set the standard for the industry, and be widely imitated, is a great lever.

4. Posterity

Similar to the effect of power, and one of the results which power helps one achieve, is fame and reputation in the future. It is strange how powerful a motive this is in human affairs, because future fame does very little for present pleasure in a physical sense. But people do care very much about what the future will think of them, and the appeal to future reputation is a very powerful lever. You could say, for instance, that the implementation of the new chemical process will be the one which will be remembered for many years in the industry as a great step forward in the technology. All those associated with this development will achieve fame in the industry, and lasting reputation with future generations of chemical engineers. It is a powerful incentive! Good name, now and in the future, is something most people will work hard to achieve. So important is it, that benefactors are prepared to pay a high price for respect from posterity.

5. Convention

As strong as power, and respect from posterity, is the appeal to convention. People like being like other people, it makes them feel safe. I suppose the general feeling is that if everyone else does it, it cannot be silly, bad, or dangerous. The majority of our actions are conventional in this sense; it is partly a love of unity, of feeling like our fellow human beings, it is partly a feeling that our individual decisions are unlikely to be as knowledgeable and experienced as those arrived at by many other people working together, and it is partly fear. If everyone is doing something, maybe you will miss out if you don't do it too.

This argument is perhaps strongest of all in a commercial and industrial context. If you can show that a major competitor is about to change over to the new chemical process you are proposing, you will have moved a very powerful lever. The fact that another similar firm has taken the decision arouses complex emotions. Firstly, fear, since they may get a competitive advantage from it; secondly pride, since they can't be allowed to steal a march; and thirdly the sense of convention, since this may become the standard process if others are adopting it. 'Others do it' is probably the strongest argument of all.

Some twenty years ago Britain founded a new group of very expensive universities, and the main argument used for it was that we would be matching the Russians and the Americans per head of population. Not an educational argument, not an economic argument, but an argument of convention. If something is accepted practice, then you can use the fact that most people have a love of unity, they want to be like everyone else, as a major lever.

If the convention is opposed to you, you may think there is little hope for your proposal. But this is not so. Most people act according to several, often contradictory, conventions. The art is to show that your proposal is a convention, even though there may be other, equally powerful but opposed conventions. Psychologists point out that people very often behave, with no sense of contradiction, according to different conventions at different times. The same person can hold contradictory opinions about charity, believing at one moment that he should help the poor, and at the next moment that giving money to the lower classes will only make them ask for more, depending on the group he is with at the time. The persuader's skill lies in aligning his message with the right conventions.

6. Sentiment

Speakers who have little experience of persuasion often think that appeals to sentiment are the most powerful appeals. They aren't. They certainly have an effect, and if you can show that your proposal will be nice for everyone else, especially dogs, cats, children and old people, it will undoubtedly be an extra persuasion. But it is a relatively weak lever, and you should not rely too much on people's altruism. It operates if all else is equal, but the appeals to convention, posterity, and power are much more effective instruments. If you are trying to persuade a local government meeting to approve a start on building a village hall, the argument that the members of the committee will be gratefully remembered by many generations of villagers will carry more weight than the argument that it will be nice for the village children to have a meeting place. People *do* like to be nice; but mainly because it gives them a sense of power.

7. Contrary suggestion

I have suggested that strong emotional appeals, which play on sentiment are weak persuasive levers. But there is one emotion which is easy to arouse, powerful in its effects, and therefore a good tool for persuasion. That is the emotion of dislike, fear and loathing. If you can create disgust at the alternatives, you are half way to winning support for your solution. The tactic is to dismiss other solutions, so the prospect looks bleak, and the problem insurmountable, unless your proposal is accepted.

Contrary suggestion works best if you do not directly run down the opposing case. People expect you to see the opposition in the worst light. A more effective tactic is to use value loaded words. Make your audience believe that there is no decent alternative. While you will only create opposition if you go too far in denigrating another proposal, you can often subtly reduce the credibility of a contrary proposition. The art is to cast a shadow over it, to suggest doubts, difficulties, and fears of an unsettling kind. Clearly, the alternative chemical process may have good arguments for it, but it may also have a doubtful safety record, and some anecdotal information about accidents may be quite enough to create an unsettling impression. If at the same time you attach the positive virtues of convention, and reputation to your scheme, it is likely to succeed.

The art, as in all persuasion, is to show how your scheme relates to the positive values. Creativity in persuasion lies in good ways of aligning the scheme you are proposing with the great human values. For example, imagine you want to persuade your company to spend as much on sports facilities for the professional and managerial staff, as it already does for the manual workers' sports. You can argue that the efficiency and health of the most important staff matters, you can argue that the new sports facilities will be an impressive legacy for the future, and you can argue that most other companies have facilities for managerial sports. But one of the most powerful arguments might be to suggest that a company which only provides facilities for manual workers is doing little more than show its fear of the unions. Good professional staff may not apply to a company which has no sports facilities for them, because they find it discriminatory. Finally you could say that the company is encouraging a race of overweight, pasty faced, and unhealthy executives. This is all contrary suggestion, and it adds greatly to the power of the message.

One of the most effective uses of contrary suggestion, which I also owe to my colleague, Peter Hunt, was an advert by the Health Education Council:

> This is what happens when a fly lands on your food. Flies can't eat solid food, so to soften it up, they vomit on it. Then they stamp the vomit in until it's a liquid, usually stamping in a few germs for good measure. Then when its good and runny, they suck it all back again, probably dropping some excrement at the same time. And then, when they've finished, it's your turn.[7]

By making the alternative disgusting, the Council puts power behind its message. That is contrary persuasion.

Advocacy

When persuading an audience, be acutely sensitive to the way they perceive you. Bias, self-interest, even dishonesty, are such permanent features of human behaviour, that an audience is always ready to suspect the speaker of lying for profit. Whenever someone shows warm partiality for an idea, the first question that is in the audience's mind is whether the speaker is an advocate or pleading for himself. This is a major issue, and absolute clarity is essential. To seem to gloss over the issue will only increase suspicion. Tell the audience as soon as possible exactly what your relation to the proposal is, otherwise suspicion will interfere with their belief in everything you say.

Such honesty is not always easy. It often takes some insight to be fully aware of your own self-interest. You may disguise it from yourself, as well as others, by wishful thinking. But the audience will have no such delusions. The average listener is acutely sensitive to the speaker's self-interest. It is very difficult for a speaker to disguise this interest from an audience. As I have already pointed out, non-verbal communication is a potent signalling mechanism, and the signals of insincerity are unmistakable.

If you have no over-riding personal interest in the scheme, you have one good advantage. You can approach the topic in the guise of a professional advocate. You can appear to deal with objections in an objective way, speaking as a servant of the audience, rather than as a servant of the cause. They will be more inclined to believe you if they think you are on their side, helping them to arrive at a fair and rational decision.

You have additional responsibilities if you want to appear as an advocate for a cause. An advocate must be well informed, and must have done his home-work. A mistaken fact will undermine your credibility, and the audience may not believe what you say thereafter. But the responsibility for accuracy is balanced by a great advantage. If the audience will accept you as an advocate, you can appear as an independent, but be persuaded and enthusiastic yourself. Over-enthusiasm for something which is to your own advantage is not an attractive quality. But enthusiasm for something where we are independent is strongly persuasive. It is important to remember, though, that advocates depend almost entirely on their credibility as communicators. It is surprising how little nature changes over the centuries, and the need to be seen as a decent and honest person was as important in ancient Greece as it is today. Aristotle wrote:

> Persuasion is achieved by the speaker's personal character when the speech is so spoken as to make us think him credible. We believe good men more fully and more readily than others: This is true generally whatever the question is, and absolutely true where exact certainty is impossible and opinions are divided. . . . It is not true, as some writers assume in their treatises on rhetoric, that the

personal goodness revealed by the speaker contributes nothing to his power of persuasion: on the contrary his character may almost be called the most effective means of persuasion he possesses.[8]

Erwin Bettinghaus concludes that:

> The Greeks referred to the *ethos* that the speaker had. Early theologians talked about charisma. Management trainers and small-group theorists have talked about leadership. Other terms which have seen wide application include image, status, prestige, and source credibility. Regardless of the term used, the variable being referred to is the same.[9]

Advocacy, in other words, depends in the end on credibility, and credibility depends on the audience perceiving you as worth believing. On the other hand, if you are pleading in your own person, you must also be liked by the audience, for they must want to do what you recommend, as well as thinking it best. I suggested earlier in the book the approach, attitudes, and behaviour which are most likeable. As with all the other skills of speaking, these techniques become especially important when trying to persuade.

There is one warning, though, for those who appear in their own person. The personal approach requires you to control your enthusiasm and put it in perspective; covert personal advantage is always suspected if you are too warm. Nor must the talk be a monument to your ego. You must be reserved if you are presenting a case which is to your own advantage. As in all speaking, it is necessary to be concrete, not theoretical, when persuading people. Only emotions will motivate, and few people feel much emotion about theory, or abstract ideas. Abstract ideas lack the immediacy which allows feelings to become involved. When their judgement of a person is involved, as well as factual issues, it is often easier for people to listen. You should not offer too many facts in a persuasive presentation. The art is to choose a limited amount of factual material – the more pointed and central the better – and restrict the quantity of your information.

Dealing with objections

The only way to win at the bridge table is to deduce what the other hands contain. Similarly, the only way to win at the persuasion game is to work out what objections can be made against you *before* you present your case. Research on the effects of different ways of dealing with objections offers interesting results for the persuasive speaker. All the work shows that the arguments of the opposition must be dealt with very carefully. If you deal with the case for the other side, giving a balanced two-sided message, you will immunize the audience against later persuasion from the opposite case. Lumsdaine and Janis proved this as long ago as 1953. They got university students to listen to one-sided, and two-sided radio lectures (which were contrived for the purpose) on the production of atomic bombs in Russia. Both the one-sided and the two-sided versions affected the students' opinions significantly. The students then listened to the opposite view. People who had heard only the one-sided message were less sure afterwards. The swing towards the opinion expressed in the first lecture dropped from about 60% to 2% after hearing the second lecture. But the people who had originally heard a lecture which gave both sides had been innoculated against the arguments from the second lecture. They remained persuaded; about 60% adopted the opinions of the first lecture.[10]

Presenting a two-sided message, which deals with the opposing case thus loses little, and the potential gains are considerable. It is therefore the preferable tactic. An important first task in preparing a persuasive case is to work out a list of the main objections that can be made. Think carefully what you would say if your role was to defeat the case. Imagine, if you like, your case being presented by your most hated colleague, and work through the objections you could find to knock him down. Imagine the most hostile thing which can be said against you, by the worst and most unreasonable opponent. Even better, if you know someone who opposes your proposal, and can reasonably talk to him before giving your presentation, do so. He will tell you what his objections are, and you can be prepared to deal with them.

When you have worked out all the objections, you may feel like abandoning your case and deserting to the other side. But don't despair; you have a good chance of winning now. You have several very real advantages, the first of which is that you can no longer be surprised by an unexpected point which has not occurred to you. You can now dodge the most pointed objections, by simply not mentioning that aspect of the topic at all. Even if there are things to be said on your side of the case, you might do better to forego those points in the hope that the objections will also be overlooked. And you can now deal in advance with the other objections. They lose much of their force if the proponent of the case has already mentioned them.

It is wise to admit defects in your case openly to your audience. On the one hand they will respect a balanced view, and on the other hand you can put the case for the opposing argument weakly. It also gives you a chance to counter them, while presenting them. But perhaps the most important advantage of mentioning the objections yourself is that it removes their novelty value for the objector, who will otherwise be credited with a 'good idea'. It also makes you seem thorough, balanced, and impartial.

An objector does not have to be able to offer an alternative solution. He merely has to point out flaws in your solution, to turn the audience against accepting your proposals. It can help to defuse this type of objector if you counter by asking for his positive proposals. They are often weak, and ill-considered, and contrary-suggestion will demolish them. You can then re-assert your own proposals as the only feasible solution.

Despite the need to demolish the objector's case, it is wise to treat objectors fairly and courteously. Not only is it good manners, it will also ensure that you do not unwittingly offend people who may have strong support among the audience. If you show tact and skill in dismantling the objection, while remaining friendly towards the objector, you will gain the admiration of the audience. This will increase your credibility, and with it your chances of success.

Ask for it

I said at the beginning of this chapter that the most common reason for the failure of persuasive talks is that speakers fail to ask for what they want. The point is so important that it is worth repeating. This repetition when starting, and when finishing, is exactly what the persuasive speaker must do. Most talks, except the very controversial ones, should start with a clear statement of what the proposal is. All talks should end with a positive request for approval. When persuasive talks fail, you will often discover that listeners are uncertain about what was wanted, had no clear call to action, and were in the dark about the real issue.

The solution is to keep the issue broad and simple. If the audience is at all confused it will take refuge in inaction. The final part of a per-suasive talk should be a summary of the arguments. It is important to ask for only one thing, even if it is a plan which includes several actions. It is also important to use a few strong arguments; don't indulge in over-kill. If you have too many arguments, the really telling ones will be lost amongst the trivia. Finally, it is also important to make the structure of the argument *very* clear. What seems simple to you, who are familiar with the material, will be complex to the audience who are new to it. What seems clear to them may well seem trivial and simplistic to you. My major conclusion, after many years of dealing with both spoken and written technical presentation, is that writers and speakers should go one stage further towards simplicity than they judge is needed. Especially in persuasive talking, it is difficult to be too clear. Err on the side of over-clarity, if you must err, never on the side of incoherence. You will gain nothing if they don't understand what you want. Don't be side-tracked, either by questions, or by your own red-herrings. Make the centre of the talk a simple, clear, request for a positive action.

One final idea which comes from psychological research: it can be useful, during discussion and questions time, to ask people individually if they agree with the proposal. You can even go so far as to ask if they will vote for it. Providing such direct questions are tactful, they will help to obtain positive action. The best way to ask people is by making a request for help at a personal level. Thus, you might say: 'Can you help me to judge if I have provided enough information about this project, by telling me if you personally are likely to support it . . .' Psychologists found that people are persuaded most when they take part in discussion in this way. One study showed that housewives could be persuaded by discussion to say they would try a new food product. Once they had committed themselves to do so, approximately a third actually did. But lectures alone had very little effect.[11]

In a similar experiment Cohen showed that if someone makes a public statement about an issue, he is much less likely to change his mind. Cohen showed that this effect doesn't operate if the person simply fills in an anonymous questionnaire.[12]

The conclusion, then, is that you must *ask* clearly, and unmistakably, for what you want. Put the audience in the position of having to say unequivocally 'no' if they want to turn down your request. In this way you use their natural inertia to get agreement.

Of course, even if you employ all these tactics, you will not always succeed. But all is not lost. There is plenty of evidence that a tactful presentation, which draws lightly and skillfully on the various techniques described in this chapter, will have a better long term effect than a presentation which tries too hard. To use the levers of persuasion coarsely, insensitively, and brashly, can do lasting damage to your case and to your own credibility. But if you have presented a skillful case you may still succeed. If your proposal is refused at the first meeting, it does not mean that you have entirely failed. The immediate effects may be disappointing, but after a while the message may have increasing effect, and eventually your case will become accepted.

Notes to chapter twelve

1. Erwin Bettinghaus, *Persuasive Communication* (Holt, Rinehart and Winston, 1980), p.1.
2. Svenn Lindskold and James Tedeschi, Self-esteem and sex as factors affecting influencability, *British Journal of Social and Clinical Psychology*, Vol.10 (1971), pp.114–122.
3. W.P. Robinson, *Language and Social Behaviour* (Penguin, 1974), p.113.
4. Alfred North Whitehead.
5. Erwin Bettinghaus, *Persuasive Communication* (Holt, Rinehart and Winston), p.28.
6. From an advertisement for *Literary Heritage Books*.
7. From a Health Education Poster, collected by Dr. Peter Hunt.
8. W.D. Ross (ed.), *The Works of Aristotle*, Vol.11, p.7.
9. Erwin Bettinghaus, *Persuasive Communication* (Holt, Rinehart and Winston), p.102.
10. Lumsdaine, A.A., and Janis, I., Resistance to Counter-propaganda, Produced by One-sided and Two-sided Propaganda Presentations, *Public Opinion Quarterly*, Vol.17 (1953), pp.311–318.
11. Lewin, K., Group Discussion and Social Change, in T.M. Newcomb and E.L. Hartley (eds.), *Readings in Social Psychology* (Holt, Rinehart and Winston, 1947).
12. Cohen, A.R., *Attitude Change and Social Influence* (Basic Books, 1964).

Further reading

Persuasion is an important subject, and a great deal has been written about it. Here is a copious selection of the recent books:

Anderson, Kenneth F., *Persuasion: Theory and Practice* Allyn and Bacon. 1978).

Bradley, Bert E., *Fundamentals of Speech Communication: the Credibility of Ideas* (W.C. Brown, 1981).

Brembeck, Winston L., and Howell, William S., *Persuasion: A Means of Social Influence* (Prentice Hall, 2nd Edn, 1976).

Brown, Hedy, and Murphy, Jeannette, *Persuasion and Coercion* (Open University Press, 1976).

Dineen, Jacqueline, *Talking Your Way To Success: The Persuasive Power Of Words* (Thorsons, 1977).

Kitzhaber, A.R., (ed.), *Persuasion and Pattern: Concepts in Communication* (Holt, Rinehart & Winston, 1980).

Larson, Charles V., *Persuasion: Reception & Responsibility* (Wadsworth, 1983, 3rd edn).

Napley, David, *The Technique of Persuasion* (Sweet and Maxwell, 1983).

Nuttin, Jozef M., *The Illusion of Attitude Change Towards a Response Contagion Theory of Persuasion* (Academic Press, 1975).

O'Donnell, Victoria, and Kable, June, *Persuasion: An Interactive Dependency Approach* (Penguin, 1982).

Roloff, Michael E. and Miller, Gerald R. (eds.), *Persuasion: New Directions in Theory and Research* (Sage Publications, 1980).

Ross, Raymond S., *Persuasion: Communication and Interpersonal Relations* (Prentice Hall, 1974).

Sandell, Rolf, *Linguistic Style and Persuasion* (Academic Press, 1977).

Simons, Herbert W., *Persuasion: Understanding, Practice & Analysis* (Speech Communication Series, Addison-Wesley, 1976).

Smart, J.J.C., *Ethics, Persuasion and Truth* (Routledge and Kegan Paul, 1984).

Thompson, Wayne Noel, *The Process of Persuasion Principles and Readings* (Harper and Row, 1975).

SUMMARY SHEET

Chapter twelve – Persuasion

Persuasion is an honourable skill, like marketing.

Either ask first, then explain; or work up to a request.

Make what you want quite clear.

You can't bore the audience into agreement.

The speaker's credibility determines his or her persuasive power.

Be clear and honest about your own motives.

The history of the organization and the topics are important for the choice of tactics.

People are motivated, in this order, by:
— habit
— emotion
— intellect.

The levers of persuasion are:
— the desire for self-preservation
— the pride in possessions
— the sense of power
— good name with posterity
— convention, and what others are doing
— sentimental care for the young and weak
— disgust at the alternatives, or contrary suggestion.

Be an honest, but accurate and committed, advocate.

Deal with objections in advance, so defusing them.

Always ask for what you want clearly, at the end.

13

Question time: leading a group discussion and answering questions

Being a chairperson

Question time is testing time. It is the audience's chance to solve the puzzles left in their minds at the end of your talk, as well as their chance to satisfy themselves that you really do know what you're talking about. So it is testing, both for you and and for them. It is also the last thing which happens during a presentation, so this is the time when the final impressions are made. Being professional in your handling of question time is thus especially important.

Your function when handling questions is, in effect, to hold a group discussion. You are, in other words, in the same position as the chairperson in a meeting. So the speaker is on his or her toes in a new way during question time, for he or she is now fully exposed, and open to probing, as well as approval for his social skills. He or she must lead the group, answer the points which are being raised, watch for people who want to ask the next question, and tactfully stop those who want to interrupt, all at the same time. It requires presence of mind, and a skilful technique, to do all this successfully. But if you can round off your presentation with a controlled and useful question-time your presentation will be even more impressive and you will gain in respect as a speaker.

The four most common reasons for disaster during question time are these:

1. failing to repeat the question so that the rest of the audience can hear it;

2. answering the wrong question, and failing to see the point of the question that *was* asked;

3. talking for half as long as the original talk in answer to one question, and almost giving a new talk when everyone else is ready to go home;

4. failing to be fair to people in the queue to ask questions, and letting the discussion become a riot with everyone talking at once.

How can you avoid this happening? By being aware of the dynamics of group discussion. This is not the terrifying task, for professional psychiatrists only, which it might seem. In practice, most people have pretty well-developed social skills, although the skills are often unconscious. But the trouble with speaking is that nerves seem to obliterate these social skills, and otherwise tactful people can find themselves blundering around on the sensibilities of their audience. As with everything else in speaking, the moral is to *think* about what you are doing. Knowledge brings confidence, correct actions practice skills, and soon proficient question handling becomes a habit. This is the basic message of the whole book, as you are by now tired of being reminded, but it's importance has not diminished.

The motivations of the people who ask questions are interesting. In a book called *Interaction Process Analysis*, Robert Bales offered a detailed breakdown of what he called the types of 'interactional moves'. These moves are the complex social manoeuvres people indulge in, to satisfy their deeper motives, while still appearing to do something socially acceptable. Bales distinguished twelve categories, ranging between the two poles of what he called the 'social-emotive' moves. At one end the moves attempt to show solidarity, and at the other to show antagonism. Moves designed to help forward the task itself (such as suggestions or ways forward) lie in the centre.[1] Bales then made use of this analysis to describe the way a group of people behave during discussion. He discovered that up to a third of the verbal contributions in a typical discussion will be concerned with controlling, manipulating, pressuring or offending others.

A large proportion of our total speech activity is said in order to massage the social aspects of our interaction with others. So do not expect the questions you receive at the end of your talk to be only genuine requests for information! Many of them will be making political, personal, or polemical points. Many of them, too, may be designed to get attention for the question asker, rather than to get information. Recognizing these facts of human interaction does not mean that you have a missionary duty to crush people who misuse question time. In most cases there will be a factual question mixed up in the self aggrandisement. And even if everyone else can see that there is nothing but self display, you will gain credit if you deal with the question courteously.

Tact is shown by forebearance, and smoothing social problems, not by highlighting them. The best solution is to answer the obviously self-advertising question with a few brief, strictly factual, sentences, and then immediately invite the next questioner. In this way you show that you too recognize the fatuity of the question (by answering it shortly), you show your own social skills by the adept way in which you have side-stepped a possible confrontation, and you change the subject. If you can achieve all this with one manoeuvre, you will gain respect from everyone else in the audience.

Basic tactics

Some speakers like to encourage questions throughout their talk. The advantage is that it creates a relaxed, homely, atmosphere in which everyone feels free to speak. The disadvantage is that the talk *nearly always* becomes side-tracked. Usually, sooner or later, one of the listeners is pricked by something which at the time seems important, but on reflection might not matter. Usually these are points which would have been answered later in the talk anyway. Because the interrupter jumps in there and then, the discussion usually veers off on a digression. I have seen *very few* of these free-for-all talks finish their topics. By all means run free discussion if there are enough well-disciplined thinkers in your group to make the discussion fruitful. But don't come with a carefully prepared list of points you want to make; you won't get through them. If you have got a set of points which you need, and want, to make, then control the discussion.

Tell the audience right at the beginning that you will accept questions at the end of the talk. If someone does interrupt you, and it is not simply a matter of a 'yes' or 'no' answer (or very nearly so) then remind them politely that you will deal with questions at the end. Usually such interruption will only happen once or twice, if you postpone them in this way. It helps to make a note of the point raised, so that you don't forget to deal with it at the end. One very impressive presentation I watched, had an exciting new subject to get across, and quite a few interruptions from the audience. The speaker made a one word note on the corner of the blackboard when each point was raised, and promised to come back to it Two things were especially well managed: firstly, these notes were made at 90° to the rest of the writing on the board, so that they stood out. Secondly, none of the points were forgotten, and as each one was dealt with, the note was erased. By the end of the talk there was a clean board, and a very impressed audience.

Trying to control questions only works if you make it quite clear what you are planning to do. If a questioner doesn't understand that you are intending to deal with questions all together at the end, he will think that his question has been ignored, not saved. He will think you have shut him up, rather than delayed him. Other members of the audience may feel the same, and the atmosphere will chill. One reason why the technique of noting questions on the board works is that people *know* they will not be forgotten. Make it clear, every time you refuse to take a question during the talk, that you will deal with it at the end, and do it in a friendly way.

When the time comes, be welcoming to the questioners. Regard them as opportunities to present your information even more clearly, as chances to clarify and repeat important points. Question time is, in fact, an opportunity for you to discover how much the audience has understood of what you've said. Right at the centre of the art of question answering is the ability to take a positive attitude to it. If a speaker is apprehensive about question time, he will probably respond defensively, and may even misunderstand a simple request for information as a personal attack. If you think of question time as having many advantages, as a chance to display tact as well as knowledge, you are likely to create a more positive, welcoming, impression at the end of your talk.

It is very easy to think that you know what you are being asked within the first few seconds of the questioner opening his mouth. You may do. But at least as often a speaker is wrong about this. There is a very simple reason: the speaker knows what he hasn't made clear, he knows the gaps and insecurities in his material, and he is nervous about what he will be challenged about. He is also very familiar with his *own* interpretation of his subject. A questioner, on the other hand, may have a completely different approach, or insight, which he wants to discuss with the speaker. All too often, it is clear to everyone else that the speaker has rushed in, and answered the question he was expecting, rather than the question that was actually asked. Wait until the asker of the question has finished speaking, listen carefully to what is said, and then summarize the question before answering it.

Summarize the question first

Summarizing the question is so important that it must be singled out for separate discussion. I have already said that failing to summarize the question is one of the most common reasons for failure at question time. A summary does not have to be a full repetition of the question. It does have to clarify what was asked. There are three powerful reasons for making a simple summary of every question, before answering it:

1. the questioner is reassured that the speaker understands his question;
2. the audience can all hear clearly;
3. it focuses the speaker's attention on the exact question.

In other words, making a summary makes sure that you have got it right, it reassures the questioner that you have grasped the point he is making, and it repeats the question for the audience so they also grasp it. Above all, it gives *you* time to think!

By a summary, I mean little more than the key phrase or topic. One useful technique for dealing with long, rambling questions is to reduce them to a few words. An impressive sense of relevance is created when a long question, consisting of several red herrings, is reduced by the speaker to the simple statement, like: 'I am asked about the cost-efficiency of the new process'. If the question is a long one it is especially important to repeat it in summary before answering it, because you have to remind the rest of the audience, who will not have been listening as attentively as you have, what the question was *really* about. The technique of a few words of summary before launching into an answer is probably the best single tip I can give a speaker about question time.

Another useful result of summarizing is that enables you to clarify what the real issue is. It is very common for a question to deal with several different issues; or it may offer a different perspective which cuts across your own distinctions. Summarizing provides a simple mechanism for dealing with this. If there are several questions wrapped into one, enumerate them when you summarize, and then take them carefully in order. In this way you will clarify what may be tangled and complex issues for the audience, and help to maintain your own clarity of mind.

Look interested

Give positive feedback to the question asker. Don't appear impatient or irritated, or seem to think him slow, imperceptive, or stupid. Look interested and enthusiastic, and try to give non-verbal messages of encouragement. It is too easy to give the *wrong* messages without meaning to. I have discussed the importance of giving the right non-verbal signals earlier in the book. At question time, the situation changes. You are now interacting with one person only during the question, and then switching to answer the question to the whole audience. You must be aware of the changes of attention. While listening to the question, pay attention closely to the questioner, look at him, smile and nod encouragement and understanding, just as you would in a normal one to one interaction.

Imagine yourself, briefly, alone in the room with the questioner. He or she is now the focus of your interest and attention, and the normal rules of courtesy insist that you appear interested. You may well be tired by the end of the talk, and it is obvious that many speakers regard question time as a winding down period when the strain is over. I have seen speakers putting on a flaccid, immobile, bored expression while looking round the rest of the audience, during a question. It is possible that all the speaker was doing was watching for the next question, while allowing the evidence of a very natural tiredness to show on his face, but it doesn't look like that to the questioner. He thinks the look implies boredom and lack of interest in the point he is making. Remember that for most people it takes quite a lot of courage to ask a question. At least do the courtesy of paying attention and looking interested.

Question time is *not* an opportunity to relax, the sign that the presentation is over, and that no more effort is needed. It *is* an opportunity to reinforce the impression of competence you should have created, an opportunity to clarify exactly what you have said, and to explain why you approached the subject in that particular way. There are fewer problems in keeping the audience's attention during question time: the variety of different voices usually does that. But there is a different problem of crowd-control. Effective control of questions and their answers is a very impressive and reassuring conclusion to a presentation.

Results from psychology can help us to understand how to achieve the effect we want. You, as speaker, are interacting with the questioner one to one during the question. Create a dialogue between yourself and question asker, in which you clarify what his point was. This is really a conversation, but it is a conversation with one difference. A whole group of other people are watching and listening. Therefore it is especially difficult to maintain a smooth interaction. Psychologists have found that it is principally eye-contact which controls the flow of conversation between people.

> In the Kendon study it was found that the terminal glance conveyed information to the other, that the speaker was about to stop speaking; if this glance were omitted, a long pause followed.[2]

> An extended look from the speaker was followed by an immediate response from the listener in 70 per cent of the cases, while not looking gave pauses or failures to respond in 71 per cent of the cases.[3]

> In Britain a nod gives the other permission to carry on talking, whereas a rapid succession of nods indicates that the nodder wants to speak himself.[4]

We usually operate this non-verbal signalling mechanism without thought, and it usually operates smoothly. But in front of an audience, it is often not so easy. If you remember to look at the questioner as you finish your answer, you will give him a sense that he can continue the interaction, and that you have finished your contribution. He or she may then nod, express satisfaction, or clarify what he was trying to ask. You will also find that if a question is going on for too long, changing from the occasional nod, to a rapid series of nods, will unconsciously tell the questioner that you think it is your turn to speak. I am not suggesting, of course, that the speaker must become an actor, who can manipulate his audience's responses like Houdini could manipulate locks. These are simple points, but like all advice about non-verbal signals, you will find that only a little conscious knowledge is required to control what you do.

Stopping sub-committees

One risk during discussion is the fragmentation of the audience into smaller groups. If there is a lot of interest, people may start talking excitedly amongst themselves: they may even start a discussion about something quite irrelevant. Whatever the subject of these small pools of discussion amongst neighbours, they threaten the successful conclusion of a presentation. You cannot stop the audience talking among itself. Human beings find it painful to listen to someone else talking for a long period. Some people even hold semi-conscious imaginary conversations in their heads, while listening. When the talk ends, it is like taking the lid off a pressure vessel. Everyone has something to say to their neighbour. You can't stop this natural reaction, and there may well be important messages people have to pass to each other.

My technique is to ignore the buzz of conversation, for a minute, but then firmly to collect the audience's attention again. People will still whisper to each other from time to time; providing this is not loud, it does little harm. The fact that it is a whisper it is at least a recognition of your prior claim to the audience's attention. But if these whispers get louder, or continuous, you should intervene. If you don't the buzz of repressed conversations will drown anything else you want to say.

There seems to be a 'critical number' for independent conversations in groups. If there are six to eight people, everyone can hear what is being said. My technique to draw these side conversations into the main discussion is to react as if the point was made to the whole group, rather than just to the neighbour. Thus, if a member of a small group remarks to his neighbour that the scheme just presented is all very well, but is very costly, I would overhear, and treat it as a question.[5]

Once the group gets much bigger than ten people, it is possible for independent conversations to start. Two techniques I like, are the joke: 'would the sub-committee in the corner like to report its findings to the main committee!' and the appeal: 'would you like to share your points with the rest of us'. One particularly powerful speaker caused a roar of laughter when he shouted: 'If it's worth hearing, tell us all, if not, shut up!' He was an unusually ebullient and affable man, in a position of great power, and I certainly don't recommend his technique to everyone. But the anecdote does underline how important it is to avoid the group fragmenting.

It also underlines the basic rationale of all group control during a presentation. One appeals to solidarity with the rest of the group. The common purpose of the group is to listen. The techniques for controlling discussion with: 'report . . . to the main committee', 'share . . . with us', and 'tell us all' are appeals which underline the point that the individual conversation is offending the *group*, not just the speaker. You would find a sullen hostility if you merely demanded that they listen to you; appeals to the group's aims always result in an apologetic silence.

Controlling conflicts

Any group is fraught with complex interactions between individuals, and currents of personal feelings. One of the strongest of these is the fear of the individual member of the audience that he or she is not being valued. Many individuals will choose to remain silent, rather than brave the searchlight of the whole group's attention which would be turned on them should they open their mouths. Those who *do* brave this limelight, expect recognition. Nothing raises hostility and resentment more quickly than having a question ignored. Therefore, the speaker must be especially careful to take note of everyone who asks a question, and deal with them, as an individual and with care.

You may find yourself in the position of having to choose between several questions at once. The usual pattern in an uncontrolled question time is for both people trying to ask questions to talk together for a few seconds. Then the louder, the strongest personality, or simply the nearest wins, and continues his question. The other person is squashed, and feels humiliated. All too often when the first question is dealt with, another person jumps in, and is allowed to get the speaker's attention, and so it goes on. The situation is unfair, and it creates an uncomfortable atmosphere.

The solution is simple. You, as speaker, are responsible for maintaining a fair distribution of questions. You must scan the audience when questions are invited, and make a mental note of people who want to be in the queue. It is usually easy enough to see: most people lean forward, make coughing sounds, or produce other gestures a few seconds before they speak. I find that if it becomes clear to my audience that I am going to run a well organized question time, people make gestures, such as half raising a hand, much like bidders at an auction. Once I have seen them I usually nod to them, to indicate I have mentally recorded their request to ask a question. It is then easy to indicate by a hand gesture whose turn is next.

If you find two people trying to talk at once, despite this attempt to be fair to everyone, it is up to you as speaker to make a choice. It is best done by a simple conventional phrase, such as 'Perhaps we can have this person first, and I will take your question next' But you *must* remember to take that question next. If you don't, you will have lost credibility, and your ability to control the audience will evaporate. Anything is better than the law of the jungle, but the audience as a whole is acutely sensitive to what is seen to be just.[6]

People do get surprisingly irritated during a heated discussion if they can't make themselves heard. Abercrombie reports that one student told him after a boisterous discussion period, when everyone was interrupting each other: 'You are the only polite one, you're the only one who stops talking when you're interrupted.'[7](7)

The content of answers

So far, I have been talking only about technique, not about content. You will have gathered by now that I think technique is just as important as content, and that a well researched presentation, full of sound information and interesting ideas, can fail miserably because of bad technique. On the other hand, a vapid meaningless talk can be a resounding success if presented in the right way. I don't particularly like this fact, since I tend myself to be an earnest rather than a sparkling speaker, but it is a fact none the less. As with most of the facts of life, it is better to accept it than to fight it. But this does not mean that content can be ignored. The perfect talk is the presentation of something the listeners find interesting and important, in a way they find entertaining. So I have dealt first with the techniques of question time. Let me now say something about content.

Many speakers are worried that they will not be asked questions, and indeed a question time which begins with uncomfortable silence is disappointing for the speaker. One way of avoiding silence at question time is to leave obvious gaps in the material, which will almost force the listeners to inquire. In this way, it is possible to 'set up' questions, although it is not a technique to recommend, since the listeners may think you have not thought about the point until then. One excellent persuasive speech I remember left out the important question of where the new building was going to be located. When asked, the speaker broke into a broad, appealing grin and said 'I hoped someone would ask that!' He then made it clear that he had thought of various possibilities, none of which was ideal, and hoped the audience would act as a committee, and arrive at some agreement on the best position.

In some situations, written questions will be prepared, and given to you in advance. Although this technique seems very formal, it may be the only way to run an effective question time after a conference paper with some hundred people in the audience. Clearly, spontaneous questions are better; they show the immediate reactions of the listeners, and the speaker's answer is fresher. But written questions let you select the ones you want to answer, and prepare what you are going to say with rather more thought. You can also make some up to suit yourself!

If no questions are forthcoming, and you feel discussion would be useful, you can invent your own questions. But do not be in too much of a hurry to start talking again. My college tutor, who spent many years lecturing to extra-mural evening-classes, told me that he had evolved a technique which rarely failed. After inviting questions, he would wait in expectant silence. He set himself a limit of three minutes by the clock. He knew when the embarrassing silence was going to end; the audience did not. He always got his question before the three minutes was over.

Most speakers step in too quickly to fill any silence, rather than leaving the audience to think over the talk, and formulate questions. Just asking, 'are there any questions?' will usually produce silence. A better way is to start discussion rolling, by asking a question of an expert in that area who happens to be in the audience. This breaks the spell of silence, and discussion may than flow freely. An audience's brains are idling in neutral, after they have been sitting listening. There is evidence that during discussion listeners become more attentive, awake, and thoughtful. Discussion, if you can get it going, is a good way to end a talk.

The awkward customer

One pitfall which every speaker will meet sometimes is the person who knows better. It is fatally easy to get involved in a childish wrangle about facts with a particularly belligerent and opinionated person in your audience. Usually these people start their contra-diction in an aggressive way, which automatically leads to a response of flat denial from the speaker, much on the lines of the 'did', 'didn't', 'did', 'didn't' game. Disputing facts with a questioner is dangerous; it becomes a haggle, and rarely achieves anything. If you say calmly, 'that is not my understanding of the situation', and if necessary go on to give your sources, the audience will doubtless believe you, not him.

It is the mark of a lively and successful talk to stir up strong reactions. Don't panic if your audience argues with you or with each other. It is both a compliment and a challenge. It is a compliment, because they are awake and interested, a result of your clear presentation of the subject. It is a challenge because an emotionally charged audience is less easy to control than a soporific one. Remember that nothing raises the emotional temperature more than people's fears that their position is threatened. Once one entrenched position is threatened, others will join in the attack. In no time: 'the air is loud with the sound of the grinding of axes.'[8]It is an uncomfortable position for a speaker to be in. He feels like a war correspondent, observing a battle field. It typically happens when you are an invited external speaker, coming into a closely knit organization. Their individual battles are unknown to you, and very often they have learned to live with situations which an outsider finds bizarre.

What can you do when this happens? The first rule is to avoid taking *any* dogmatic position on issues. If you have already done so, it may be tactful to show that you understand the alternative point of view, perhaps by making a quick summary of it. If the entrenched postions in the group run completely counter to your own sympathies, then at least be good humoured about it. The first resort is withdrawal. This can be done by tactful silence. If the audience is arguing between themselves, you may be best advised to keep quiet and listen. Don't try to adjudicate, or add your weight to one side or the other. If you are directly challenged, try to show your awareness of the opposite position, and explain why you arrived at the conclusion you did. If none of this works, the last resort is to say something like: 'I hope you have found it useful to hear the other point of view on this subject'.

Whichever solution you adopt, the key is to maintain good humour. Argument and opposition are a fact of life for every one: nobody minds disagreement. It is only bad temper, ill will, and heated dogmatism which people find objectionable. If you can maintain a smile, a reasonable attitude, and a cogent explanation of why you think as you do, you will gain nothing but respect.

Speakers usually underestimate the amount of sympathy their position receives from the audience. Most of the audience will react in the same way as the speaker. If you find a questioner tedious, opinionated, or rude, you may be sure that most of the audience do. You can therefore assume that you have support, and have no need to be defensive. Tact, and skilful handling are what is needed, not counter-attack. Research on group interaction shows that audiences dislike people who use question time for self advertisement, to display a chip on their shoulder, or for other unsavoury motives. Fouriezos, Hull, and Guetzkow spent their time watching 72 conferences. They recorded what appeared to be the motivations of people who asked questions, assessing how far they had self-centred motives, such as wanting to gain a point, get something off their chests, etc. The more questions of this type there were at a conference, the less likely was the meeting to get through its agenda. When they were questioned after the session, most people expressed greater dissatisfaction with the meeting when there were many such questions.[9]

What I have said so far encourages a forgiving, accepting attitude on the part of the speaker towards the awkward question. Indeed, I think this is nearly always the best path to take. But there will be times when it is plainly too passive a reaction to troublesome questioners. There are four powerful techniques which can be used to silence the difficult person. All should be used with care, for they are all planned affronts to the questioner. Only use a powerful technique when it has been plain *for some time* that the difficult interruption is wasting everyone else's time. The four techniques are:

1. to repeat or summarize awkward questions in a way which shows up their silliness;
2. when there is a long rambling speech given under the pretence of a question, to ask, 'but what is the real question?'
3. to use the politician's bluff ('does that answer your questions' when it has nothing to do with them) to talk about something useful instead of answering an irrelevant question;
4. to squash awkward or fake questions by making a very brief summary. You can also diffuse a very long and rambling questioner by simply replying, after a pause for thought, 'Yes', or 'No'. Pause again thoughtfully, and then take the next question.

Inexperienced speakers are always worried about being asked questions they can't answer. For many, it is a recurrent nightmare, and for most it is a growing anxiety as they approach the talk. Such fears are unnecessary; nobody knows everything. Inexperienced speakers, because of their prominent position, seem to think that they must pretend to omniscience: if they are not (as most of us aren't) they fudge, and end up embarrassing themselves, and the audience. Why? If you don't know the answer, say so. The credibility of what you *have* said will only be increased by frankness about what you don't know.

The embarrassed questioner

Shyness and embarrassment are common emotions, and ones which nearly always evoke sympathy and understanding from others. I doubt if there is anyone alive who has not been embarrassed at least once in their lives. Everyone knows what it feels like; no one likes it. And, like yawning, it is contagious. To be with someone who is obviously suffering extremes of embarrassment is uncomfortable.

Just as you are bound, sooner or later, to meet an awkward questioner, so you will meet an embarrassed questioner. He or she may take a lot of encouragement to open their mouths, then speak so quietly that no one can hear. If asked to speak up, he or she will go bright red, and speak nearly as quietly. The question may end abruptly, or trail off into mumbling, and it may prove impossible to get any more information. He or she may prefer to give up, rather than go on. What do you do? The solution is two fold: knowledge and sympathy. Firstly, be aware of the causes of embarrassment, and how to deal with it. Psychologists have done much research on this topic:

> Goffman (1955) offered a theory of embarrassment: people commonly present a self which is partly bogus; if this image is discredited in the course of interaction, embarrassment ensues. . . . A second source of embarrassment is 'rule-breaking'. Garfinkel (1963) has carried out some intriguing 'demonstrations' in which investigators behaved in their own homes like lodgers, treated other customers in shops as salesmen, or flagrantly broke the rules of games – such as by moving the opponent's pieces. This produced embarrassment, consternation and anger. . . . When a person is embarrassed, the others present usually want to prevent the collapse of social interaction, and will help in various ways. To begin with they will try to prevent loss of face by being tactful . . . They may pretend that nothing has happened, make excuses for the offender – he was only joking, was off form etc., or in some other way 'rescue the situation'. Finally, if face is irrevocably lost they may help the injured party to rehabilitate himself in the group in a new guise. . . . When A's face has been disbelieved or discredited there are various strategies open to him. . . . If he is rattled he may fall back on . . . 'Look here, young man, I've written more books about this subject than you've read', 'Do you realise that I . . .', etc. . . . What very often happens is that A forms a lower opinion of a person or group that does not treat him properly, and he goes off to present himself to someone else.[10]

Blushing and stammering, the signs of shyness, are not the only results of embarrassment. Sometimes they can be anger, defensive aggressiveness, or withdrawal from the situation. All these reactions are equally unfortunate from the point of view of a talk. The person will get nothing from the presentation, and the rest of the audience will be discomforted. Embarrassment is good for no one, and requires careful handling. If a questioner is embarrassed and loses face, an understanding, helpful and concerned response from the speaker is best. The speaker will gain added respect from the audience, and perhaps even the final loyalty of the embarrassed questioner. He or she will also prevent the atmosphere of the talk becoming unpleasant, and leaving bitter tasting memories with the whole audience.

One more reflection: since embarrassment is caused, on this theory, by loss of face, that is to say by being detected in pretence, only straightforward honesty will help. This is why I recommended being frank if you do not know the answer to a question. It is also why it is better to establish a close relationship with the audience, rather than hostility. If something does go wrong, a brief explanation and apology, or a simple smile, will often win back an audience who are, in any case, anxious to repair the damaged interaction. Remember, too, that asking questions places a strain on the audience itself. Research using measurements of pulse shows that not only the speaker's pulse, and the asker of the question's pulse, but the audience's pulse generally speeds up. Everyone feels a rise in tension, a slight excitement, a tightening of awareness, as a question is being asked. People sitting next to the questioner, especially, experience a rise in pulse rate.[11]

Conclusions

There are many more things which could be said about question answering. My impression is that it is often the part of a talk which goes least well, and an otherwise excellent presentation often falls apart in shambles in the last ten minutes. Strangely, most audiences don't seem to mind that much, I suppose because they expect that question time will be something to be endured, rather than enjoyed. But it needn't be. Careful control and awareness can turn question time into a useful and impressive part of the presentation. The speaker who is aware of his audience's moods, who chooses among competing questioners fairly, who helps the embarrassed, and gently controls the loquacious, will be remembered with affection. The final thing the speaker must do is to watch the clock. If you have announced ten minutes question time, don't let it become 30 minutes.

Of course, there will be more questions people would like to ask, of course the time won't be long enough, but you will please everyone who doesn't want to ask questions by letting them get away on time.

Let me go back to what I said at the beginning of this chapter. There are four reasons why question time goes wrong:

1. failing to repeat the question so that the rest of the audience can hear it;
2. answering the wrong question, and failing to see the point of the question that *was* asked;
3. talking for half as long as the original talk in answer to one question, and giving a completely new talk when everyone else is ready to go home;
4. failing to be fair to people in the queue to ask questions, and letting the discussion become a riot with everyone talking at once.

If you think carefully about these points, you will learn to be an effective speaker at question time as well as during the presentation itself.

Notes to chapter thirteen

1. Bales, R.F., *Interaction Process Analysis* (Addison Wesley, 1950).
2. Michael Argyle, *The Psychology of Interpersonal Behaviour* (4th Edn, Penguin, 1983), p.83.
3. W.P. Robinson, *Language and Social Behaviour* (Penguin, 1974), p.138.
4. John Corner and Jeremy Hawthorn, (eds), *Communication Studies: an introductory reader* (Arnold, 1980), p.55.
5. See Shuter, Robert, *Understanding Misunderstanding: Exploring Interpersonal Communication* (Harper and Row, 1979).
6. See Lewin, Kurt, (Ed. Gertrud Weiss Lewin), *Resolving Social Conflicts: Selected Papers on Group Dynamics* (London, Souvenir Press, 1973).
7. M.L.J. Abercrombie, *The Anatomy of Judgement* (Penguin, 1979), p.86.
8. W.J.H. Sprott, *Human Groups* (Penguin, 1967), p.137.
9. Fouriezos, N.T., Hull, M.L., and Guetzkow, H., Measurement and Self-orientated Needs in Discussion Groups, *Journal of Abnormal and Social Psychology*, Vol.45 (1950), pp.682–90.
10. Michael Argyle, *The Psychology of Interpersonal Behaviour*, pp. 203–5.
11. Donald Bligh, *What's the Use of Lectures?* (Penguin, 1971), p.127.

Further reading

For more understanding of behaviour in groups, try:-

Aronson, Elliot, *The Social Animal* (4th Edn, W.H. Freeman & Co., 1984).

Eiser, J. Richard (ed.), *Attitudinal Judgement* (Springer-Verlag, 1984).

Miller, George A., Grusky, Oscar (eds), *The Sociology of Organizations: Basic Studies* (New York Free Press, 1981).

Chapter thirteen – Questions

During questions, the speaker is a chairperson.

The four commonest reasons for failure are:
— not summarizing the question for everyone else
— answering the wrong question
— talking too long in answer to a question
— being unfair to people queuing to ask questions.

Many questions are for self-display, or for any reason other than information.

Random questions interrupt and side-track a talk.

Always summarize the question before answering.

Look interested and encouraging to questioners.

Control other conversations among the audience.

Administer the opportunities for questions fairly.

Be honest if you don't know the answer.

Deal fairly, but firmly, with awkward questioners.

Help embarrassed questioners by sympathy and interest.

A competent question time ends the talk impressively.

14

Conclusion

The only remaining advice I can give is to repeat what I said at the beginning. Practice will improve your skills. If you *think* about what you are doing, *know* something about audience psychology and the do's and don'ts of speaking, and *listen* to criticism and advice from friends, you can become a sound and effective speaker. Get a close and trusted friend to observe you practise your talk, and take careful note of what he or she says. When you have given the talk, come back to this book, and read the chapters relating to what *you* judge you did best, and did worst. You may find enlightenment on the reasons why things went well, or badly.

If you have to give a series of talks, or have to talk regularly, try re-reading a chapter each time you are preparing for a talk. As with most books of advice, there is, I know, too much material to sink in at a first reading. You will also not see the point of every piece of advice until you have met the situation it refers to yourself. Providing you do not relapse into self-satisfaction as a speaker, practise should help you to get better and better. Further than this, no book of advice can go. I hope you find what I have been able to offer useful. And, by the way, good luck with your talk!

Further reading

Here is a short list of recent books on speaking skills:

1. Barker, Larry, *Communication* (3rd edn, Prentice Hall, 1984).
2. Campbell, K. Kors, *The Rhetorical Act* (Wadsworth, 1982).
3. Mares, Colin, *Communication* (English Universities Press, Teach Yourself Books, 1974).
4. Ross, Raymond S., *Essentials of Speech Communication* (2nd edn, Prentice Hall, 1984.

5. Samovar, Larry A., and Mills, Jack, *Oral Communication: Message and Response* (William Brown, 1967).
6. Surles and Stanbury, *The Art of Persuasive Talking* (McGraw-Hill, 1980).
7. Verderber, Rudolph F., *Communicate!* (4th edn, Wadsworth, 1984).

A Specialist bibliography

For students, and those who want to explore the subject at a more technical level, there follows an extensive bibliography of speaking skills. You may also find it interesting if you want to see how much research support there is for the ideas presented in this book.

Abercrombie, M.L.J., *The Anatomy of Judgement* (Penguin, London, 1979).

Adams-Webber, Jack, and James C. Mancuso (eds), *Applications of Personal Construct Theory* (Academic Press, London, 1983).

Adler, R.B. *Confidence in Communication: Guide to Assertive Social Skills* (Holt, Rinehart & Winston, London 1977).

Adler, Ronald B., and Rodman, George, *Understanding Human Communication* (Holt, Rinehart and Winston, London, 1982).

Adler, Ronald B., Rosenfeld, Lawrence B., and Towne Neil, *Interplay: The Process Of Interpersonal Communication* (Holt, Rinehart and Winston, London, 1983).

Adler, Ronald B., and Towne Neil, *Looking Out/Looking In: Interpersonal Communication* (Holt, Rinehart and Winston, London, 1984).

Anderson, Kenneth F., *Persuasion: Theory and Practice* (Allyn and Bacon, Boston, Mass., 1978).

Annett, J., *Feedback and Human Behaviour* (Penguin, London, 1969).

Appel, S.S., Modifying Solo Performance Anxiety in Adult Pianists, *Journal of Music Therapy*, Vol.13(1), (1976), pp.2–16.

Argyle, M., and Dean, J., Eye-contact, Distance and Affiliation, *Sociometry*, 28 (1965), pp.289–304.

Argyle, M., and Kendon, A., The Experimental Analysis of Social Performance, *Adv. Exp. Soc. Psychol.* Vol.3 (1967), pp.55–98.

Argyle, Michael, *Social Interaction* (Methuen, London, 1969).

Argyle, M., Salter, V., Nicholson, H., Williams, M., and Burgess, P., The Communication of Inferior and Superior Attitudes by Verbal and Non-verbal Signals, *British Journal of Social and Clinical Psychology*, Vol.9 (1970), pp.222–31.

Argyle, M., Alkema, F., and Gilmour, R., The Communication of Friendly and Hostile Attitudes by Verbal and Non-verbal Signals, *European Journal of Social Psychology*, Vol. 2 (1971), pp.385–402.

Argyle, M., and McHenry, R., Do Spectacles Really Affect Judgements of Intelligence? *British Journal of Social and Clinical Psychology*, Vol.10 (1971), pp.27–9.

Argyle, M., Lefebvre, L., and Cook. M., The Meaning of Five Patterns of Gaze, *European Journal of Social Psychology*, Vol.4 (1974), pp.125–36.

Argyle, Michael, and Trower, Peter, *Person to Person: Ways of Communicating* (Harper and Row, New York, 1979).

Argyle, Michael, *The Psychology of Interpersonal Behaviour* (4th Edn., Penguin, London, 1983).

Arnold, Magda B., *Memory and the Brain* (Lawrence Erlbaum Associates, Hillsdale, N.J., 1984).

Aronson, E. and Golden, B.W., The Effect of Relevant and Irrelevant Aspects of Communicator Credibility on Opinion Change, *Journal of Personality*, Vol.30 (1962), pp.935–46.

Aronson, E., Turner J.A., and Carlsmith, J.M., Communicator Credibility and Communication Discrepancy as Determinants of Opinion Change, *Journal of Abnormal and Social Psychology*, Vol.67 (1963), pp.31–6.

Aronson, Elliot, *The Social Animal* (4th Edn., W.H. Freeman & Co., Oxford, 1984).

Asch, S.E., Forming Impressions of Personality, *Journal of Abnormal and Social Psychology*, 41 (1946), pp.258–90.

Asch, S.E., *Social Psychology* (Prentice-Hall, Englewood Cliffs, N.J., 1952).

Athos, Anthony G., and Gabarro, John J., *Interpersonal Behaviour: Communication and Understanding in Relationships* (Prentice-Hall, Englewood Cliffs, N.J., 1978).

Atkinson, Richard C. and Shiffrin, Richard M., The Control of Short-term Memory, *Scientific American* (August 1971), pp.82–90.

Baddeley, Alan D., *The Psychology of Memory* (Harper and Row, New York, 1976).

Baddeley, Alan D., *Your Memory: a User's Guide* (Sidgwick and Jackson, London, 1982).

Baird, F.J., Preparation, an Antidote for Stage Fright, *School Musician*, Vol.32 (January, 1961), pp.34–5.

Bales, Robert F., *Interaction Process Analysis* (Addison Wesley, London, 1950).

Bales, Robert F., *Personality and Interpersonal Behaviour* (Holt, Rinehart, and Winston, London, 1970).

Bales, Robert F., Cohen, Stephen, and Williamson, Stephen. *SYMLOG: a System for the Multiple Level Observance of Groups* (New York Free Press, 1979).

Barabasz, A.F., A Study of Recall and Retention of Accelerated Lecture Presentation, *Journal of Communication*, Vol.18 (1968), No.3, pp.283–7.

Barker, Larry, *Communication* (3rd edn, Prentice Hall, Englewood Cliffs, N.J., 1984).

Barker, Sarah, *The Alexander Technique* (Bantam Books, London, 1978).

Barthes, Rowland, *Mythologies* (English translation by Annette Lavers, Johnathan Cape, London, 1972).

Beard, Ruth, M., *Teaching and Learning in Higher Education* (Penguin, London, 1976).

Beattie, Geoffry, *Talk: An Analysis of Speech and Non-Verbal Behaviour in Conversation* (Open University Press, Milton Keynes, 1983).

Beckenbach, Edwin Ford, and Tompkins, Charles B., (eds) *Concepts of Communication: Interpersonal, Intrapersonal, and Mathematical* (Wiley, Chichester, 1971).

Beighley, K.C., An Experimental Study of Three Speech Variables on Listener Comprehension, *Speech Monographs*, Vol.21 (1954), pp.248–53.

Beisecker, Thomas B., and Parson, Donn W. (eds), *The Process of Social Influence: Readings In Persuasion* (Prentice Hall, Englewood Cliffs, N.J., 1972).

Berelson, B., *Content Analysis in Communication Research* (Free Press, 1952).

Berlyne, D.E., *Structure and Direction in Thinking* (Wiley, New York, 1965).

Bernstein, B.B., Linguistic Codes, Hesitation Phenomena and Intelligence, *Language and Speech*, Vol.5 (1962), pp.31–46.

Bernstein, B.B., Social Class, Linguistic Codes, and Grammatical Elements, *Language and Speech*, Vol.5, (1962), pp.221–40.

Bettinghaus, Erwin Paul, *Persuasive Communication* (Holt, Rinehart and Winston, London, 1980).

Bion, W.R., Experiences in Groups, *Human Relations*, Vols 1–4 (1948–1951).

Bligh, Donald A., *What's the Use of Lectures?* (Penguin, London, 1971).

Bloch, V., Brain Activation and Memory Consolidation, in *Neural Mechanisms of Learning*, Eds Rosenzweig and Bennett, pp.583–90.

Bormann, Ernest G., *Interpersonal Communication In The Modern Organization* (Prentice-Hall, Englewood Cliffs, N.J., 1969).

Bormann, Ernest G., and Bormann, Nancy C., *Effective Small Group Communication* (Burgess Publishing Co., Minneapolis, 1972).

Brackman, J., The Put On, *New Yorker*, 24th Jun 1967, pp.34–73.

Bradley, Bert E., *Fundamentals of Speech Communication: the Credibility of Ideas* (W.C. Brown, Dubuque, Iowa, 1981).

Brembeck, Winston L., and Howell, William S., *Persuasion: A Means of Social Influence* (Prentice Hall, Englewood Cliffs N.J. 2nd edn, 1976).

Brooks, William D., *Speech Communication* (4th edn., William C. Brown, Dubuque, Iowa, 1981).

Brown, Hedy, and Murphy, Jeannette, *Persuasion and Coercion* (Open University Press, Milton Keynes, 1976).

Brown, James A.C., *Techniques of Persuasion from Propaganda to Brainwashing* (Penguin, London, 1963).

Brown, Mark, *Memory Matters* (David and Charles, Newton Abbot, 1978).

Brown, R., *Social Psychology* (Collier-Macmillan, West Drayton, 1965).

Bruner, J.S., Goodnow, Jacqueline J., and Austin, George A., *A Study of Thinking* (Chapman and Hall, London, 1956).

Bugenthal, D., Kaswan, J.W., and Love, L.R., Perception of Contradictory Meanings Conveyed by Verbal and Non-verbal Channels, *Journal of Personal and Social Psychology*, Vol.16, (1970), pp.647–55.

Bull, Peter, *Body Movement and Interpersonal Communication* (Wiley, Chichester, 1983).

Bullowa, Margaret, (ed.), *Before Speech: the Beginning of Interpersonal Communication* (Cambridge University Press, Cambridge, 1979).

Campbell, K. Kors, *The Rhetorical Act* (Wadsworth, Belmont, C.A., 1982).

Carlson, Neil R., *Physiology of Behaviour* (2nd edn., Allyn & Bacon, Inc., Boston, MA, 1981).

Cartwright, D. and A.F. Zander, (eds) *Group Dynamics* (Tavistock Publications, London, 1955).

Cecco, John P. de, (ed.), *The Psychology of Language, Thought and Instruction* (Holt, Rinehart and Winston, London, 1969).

Chaikin, A.L., and Derlega, V.J., Self-disclosure in J.W. Thibaut, J.T. Spence and R.C. Carson (eds) *Contemporary Topics in Social Psychology* (General Learning Press, Morristown, N.J., 1976).

Cherry, Colin, *On Human Communication: A Review, a Survey, and a Criticism* (MIT Press, Cambridge, Mass., 1966).

Ching, J., *Performer and Audience, an Investigation into the Psychological Causes of Anxiety and Nervousness in Playing Singing and Speaking Before an Audience* (OUP, Oxford, 1947).

Clark, Herbert C., Clark, Eve V., *Psychology and Language: An Introduction to Psycholinquistics* (Harcourt Brace Jovanovich, Inc., New York, 1977).

Clark, G.K., and Clark, E.B., *The Art of Lecturing* (Heffers, 1957).

Coats, W.D., and Smidchens, U., Audience Recall as a Function of Speaker Dynamism, *Journal of Educational Psychology*, Vol.57 (1966), No.4, pp.189–91.

Cohen, A.R., *Attitude Change and Social Influence* (Basic Books, 1964).

Cohen, A.R., Some Implications of Self-esteem for Social Influence, in Janis, I.L., and Hovland, C.I. (eds), *Personality and Persuadability* (Yale University Press, New Haven, Conn., 1959), pp.102–21.

Collett, P., *Social Rules and Social Behaviour* (Blackwell, Oxford, 1977).

Cook, M., The Incidence of Filled Pauses in Relation to Part of Speech, *Language and Speech*, Vol.14 (1971), pp.135–40.

Cook, M., Anxiety, Speech Disturbances and Speech Rate, *British Journal of Social and Clinical Psychology*, Vol.8, (1969), pp.13–21.

Cooper, B., and Foy, J.M., Evaluating the Effectiveness of Lectures, *Universities Quarterly*, Vol.21 (1967), No.2, pp.182–85.

Corder, P. Pitt, *Introducing Applied Linquistics* (Penguin, London, 1975).

Corner, John and Hawthorn, Jeremy, (eds), *Communication Studies; An*

Introductory Reader (Edward Arnold, London, 1980).

Crawford, D.G., and Signori, E.I., An Application of the Critical Incident Technique to University Teaching, *Canadian Psychologist*, Vol.3a, No.4, (1962).

Daly, John A. and James C. McCroskey, *Avoiding Communication; Shyness, Reticence and Communication Apprehension* (Sage, Focus Edn, Beverly Hills, 1984).

Dashiell, J.F., An experimental Analysis of some Group Effects, *Journal of Abnormal Social Psychology*, Vol.25 (1930), pp.290–9.

Davitz, J.L., *The Communication of Emotional Meaning* (Greenwood Press, London, 1976).

Deutsch, M., Experimental Study of the Effects of Co-operation and Competition upon Group Processes, *Human Relations*, Vol.3 (1949), pp.191–231.

Dibner, A.S., Cue-counting: a Measure of Anxiety in Interviews, *Journal of Consulting Psychology*, Vol.20, (1956), pp.475–8.

Dibner, A.S., Ambiguity and Anxiety, *Journal of Abnormal Social Psychology* Vol.56 (1958), pp.158–74.

Dineen, Jacqueline, *Talking Your Way To Success: The Persuasive Power Of Words* (Thorsons, Wellingborough, 1977).

Douglas, Tom, *Groups: Understanding People Gathered Together* (Tavistock, London, 1983).

Druckman, Daniel, Richard M. Rozelle, and James C. Baxter, *Non-verbal Communication: Survey, Theory and Research* (Serge Library of Social Research, London, 1982).

Duncan, S.W., and Fiske, D.W., *Face to Face Interaction* (Erlbaum, Hillsdale, N.J., 1977).

Duncan, S., Non-verbal Communication, *Psychological Bulletin* Vol.72 (1969), pp.118–37.

Ebbinghaus, Hermann, *Memory* trans. by D. H. Ruyer and C. E. Bussenius (New York, 1913).

Edwards, A.D., *Language in Culture and Class* (Heinemann, London, 1976).

Eiser, J. Richard (ed.), *Attitudinal Judgement* (Springer-Verlag, London, 1984).

Ekman, P., and Friesan, W.V., The Repertoire of Non-verbal Behaviour: Categories, Origin, Usage and Coding, *Semiotica*, Vol.1 (1969), pp.49–98.

Ekman, P., Friesan, W.V., and Ellsworth, P., *Emotions in the Human Face* (Pergamon, Elmsford, N.Y., 1971).

Ekman, P., and Friesan, W.V., Non-verbal Leakage and Clues to Deception, *Psychiatry* Vol.32 (1969), pp.88–106.

Erskine, C.A., and O'Morchoe, C.C.C., Research on Teaching Methods: Its Significance for the Curriculum, *Lancet*, Vol.1 (1966), pp.709–11.

Eysenck, Hans and Michael, *Mindwatching* (Michael Joseph, Garden City, N.Y., 1981).

Faber, R.N., How we Remember what we See, *Scientific American* (May, 1970), p.105.

Fabun, Don, *Communication: The Transfer of Meaning* (Glencoe Press, Collier Macmillan, New York, 1968).

Festinger, L., Schachter, S., and Back, K., *Social Pressures in Informal Groups* (Harper, New York, 1950).

Festinger, Loen, *A Theory of Cognitive Dissonance* (Harper and Row, New York, 1957).

Festinger, L., and Maccoby, N., On Resistance to Persuasive Communications, *Journal of Abnormal Social Psychology*, Vol.68 (1964), pp.359–66.

Fouriezos, N.T., Hull, M.L., and Guetzkow, H., Measurement and Self-orientated Needs in Discussion Groups, *Journal of Abnormal and Social Psychology*, Vol.45 (1950), pp.682–90.

Fransella, Fay (ed.), *Personality: Theory, Measurement and Research* (Methuen, London, 1981).

Freyberg, P.S., The Effectiveness of Note Taking, *Education for Teaching*, (February 1956), pp.17–24.

Funkenstein, D.H., The Physiology of Fear and Anger in Coopersmith, S. (ed.), *Frontiers of Psychological Research*, Readings from Scientific American, (Freeman, Oxford, 1966).

Gale, Anthony and Chapman, Antony J., (Eds.) *Psychology and Social Problems: An Introduction to Applied Psychology* (Wiley, New York, 1984).

Garnier, W.R., *Uncertainty and Structure as Psychological Concepts* (Wiley, New York, 1972).

Gergen, Kenneth J., and Mary M. Gergen, *Social Psychology* (Harcourt, Brace, Jovanovich, New York, 1981).

Giglioli, Pier Paolo, (ed.), *Language and Social Context: Selected Readings* (Penguin Education, London, 1975).

Giglioli, Pier Paolo (ed.), *Language and Social Context* (Penguin, London, 1972).

Giles, H., Evaluative Reactions to Accents, *Educational Review*, Vol.22 (1970), pp.211–27.

Giles, Howard and Robert St. Clair. *Language and Social Psychology* (Basil Blackwell, Oxford, 1979).

Goffman, Erving, *The Presentation of Self in Everyday Life* (Penguin, London, 1971).

Goffman, Erving, *Encounters: Two Studies in the Sociology of Interaction* (Allen Lane, London, 1972).

Goffman, Erving, *Frame Analysis: an Essay on the Organization of Experience* (Harper and Row, New York, 1974).

Goffman, Erving, *Forms of Talk*, (Blackwell, Oxford, 1981).

Goffman, Erving, *Interaction Ritual: Essays on Face-to-face Behaviour* (Pantheon Books, New York, 1982).

Goffman, Erving, *Stigma: Notes on the Management of Spoiled Identity* (Penguin, London, 1963).

Goldman-Eisler, F., The Significance of Changes in the Rate of Articulation,

Land & Speech, Vol.4 (1961), pp.171–174.

Gomemiewski, R.T., *The Small Group* (University of Chicago Press, 1962).

Gordon, George, *Persuasion: Theory & Practice of Manipulative Communication* (Studies in Public Communication, Hastings, 1971).

Greene, Judith and Carolyn Hicks, *Basic Cognitive Processes* (Open University Press, Milton Keynes, 1984).

Griffitt, W., and Veitch, R., Ten Days in a Fall-out Shelter, *Sociometry*, Vol.37 (1974), pp.163–73.

Grindea, C. (ed.), *Tensions in the Performance of Music* (Kahn Averill, London, 1978).

Gronbeck, Ehninger and McKerrow, *Principles and Types of Speech Communication* (Scott Foreman, Glenview, Illinois, 1984).

Guetzkow, H. (ed.), *Groups, Leadership and Men* (Carnegie Press, 1951).

Gurnee, H., A Comparison of Collective and Individual Judgements of Fact, *Journal of Experimental Psychology*, Vol.21 (1937), pp.106–12.

Halacy, Daniel Stephen, *How to Improve your Memory* (F. Watts, London, 1977).

Hall, E.T., *The Silent Language* (Doubleday, New York, 1959).

Hall, Geoffrey, *Behaviour: an Introduction to Psychology as a Behavioural Science* (Academic Press, London, 1983).

Halle, Morris, Bresnan, Joan, and Miller, George (eds.), *Linguistic Theory and Psychological Reality* (MIT Press, Cambridge, Mass. 1978).

Halliday, M.A.K., *Language and Social Networks* (Edward Arnold, London, 1979).

Hampson, Sarah, E., *The Construction of Personality* (Routledge and Kegan Paul, London, 1982).

Hare, A.P., Interaction and Consensus in Different Sized Groups *American Sociological Review*, Vol.17 (1952), pp.61–267.

Hare, A. Paul, Borgatta, Edgar F., and Bales, Robert F. (eds.), *Small Groups: Studies in Social Interaction* (Knopf, New York, 1965).

Hargie, Owen, Christine Saunders and David Dickson, *Social Skills in Interpersonal Communication* (Croom Helm, London, 1983).

Harris, Luke, and Harris, Julian, *How To Become A Powerful And Persuasive Public Speaker* (A. Thomas, London, 1972).

Hartley, J., and Cameron, A., Some Observations on the Efficiency of Lecturing, *Educational Review*, Vol.20, No.1 (1967), pp.30–7.

Harvey, Ian, *The Technique of Persuasion: An Essay in Human Relationships* (Falcon Press, 1951).

Heintz, Ann C., *Persuasion* (Loyola, Communication Education Series, Chicago, 1974).

Henriques, Julian, Wendy Holloway, Cathy Urwin, Couze Venn, Valerie Walkerdine, *Changing the Subject, Psychology, Social Regulation, and Subjectivity* (Methuen, London, 1984).

Hinde, R.A. (ed.), *Non-Verbal Communication* (Cambridge University Press, Cambridge, 1972).

Homans, George C., *The Human Group* (Routledge and Kegan Paul, London, 1951).

Hovland, C.I., and W. Mandell, An Experimental Comparison on Conclusion Drawing by the Communicator and the Audience, *Journal of Abnormal and Social Psychology*, Vol.47 (1952), pp.581–8.

Hovland, C.I. and Janis I.L. (eds), *Persuasion and Persuadability* (Yale University Press, New Haven, Conn., 1959).

Hovland, C.I., and others, *The Order of Presentation in Persuasion* (Yale University Press, New Haven, Conn., 1961).

Hovland, Carl I., Janis, Irving L., and Kelley, Harold H., *Communication and Persuasion: Psychological Studies of Opinion Change* (Yale University Press, New Haven, Conn., 1963).

Howe, M.J.A., and J. Godfrey, *Student Note Taking as an Aid to Learning* (Exeter University Teaching Services, 1977).

Hudson, R.A., *Sociolinguistics* (Cambridge University Press, Cambridge, 1980).

Johan, William George, *Persuasive Speaking* (Elliot Right Way Books, Kingswood, 1972).

Johnson, P.E., Some Psychological Aspects of Subject Matter Structure, *Journal of Educational Psychology*, Vol.58 (1967), No.2, pp.75–83.

Jourard, S.M., *The Transparent Self* (Van Nostrand, Princeton, N.J., 1964).

Jourard, S.M., *Self-Disclosure* (Wiley, London, 1971).

Karlins, Marvin and Abelson, Herbert, *Persuasion: How Opinions and Attitudes are Changed* (Springer Verlag, New York, 1970).

Kasl, S.V., and Mahl, G.F., The Relationship of Disturbances and Hesitations in Spontaneous Speech to Anxiety, *Journal of Personal and Social Psychology*, Vol.1 (1965), pp.425–33.

Katona, G., *Organising and Memorising* (Columbia University Press, 1940).

Katz, D., and Braly, K.W., Racial Prejudice and Racial Stereotypes, *Journal of Abnormal Social Psychology*, 30 (1933), pp.175–93.

Katz, D., *Gestalt Psychology* (Ronald Press, 1952).

Kelley, H.H., The Warm-cold Variable in First Impressions of Persons, *Journal of Personality*, Vol.18 (1950), pp.431–9.

Kendon, A, Some Functions of Gaze Direction in Social Interaction, *Acta Psychologica*, Vol.26 (1967), pp 22–63. Also Vol.28(1) (1967), pp.1–47.

Kendon, A., Some Relationships Between Body Motion and Speech, in Siegman, A., and Pope, B., *Studies in Dyadic Interaction* (Pergamon, Oxford, 1970).

Kendon, A., *Studies in the Behaviour of Social Interaction* (Indiana, University, Bloomingdale, 1977).

Kitzhaber, A.R., (ed.), *Persuasion and Pattern: Concepts in Communication* (Holt, Rinehart & Winston, New York 1980).

Klauss, Rudi, and Bass, Bernard, *Interpersonal Communication in Organizations* (Academic Press, London, 1982).

Kleck, Robert, and William Nuessle, Congruence Between the Indicative and Communicative Functions of Eye Contact in Interpersonal

Relations, *British Journal of Social and Clinical Psychology*, Vol. 7 (1968), p.241, 243.

Klein, J., *Working with Groups* (Hutchinson, London, 1961).

Klein, J., *The Study of Groups* (Routledge and Kegan Paul, London, 1965).

Kline, Paul, *Personality Measurement and Theory* (Hutchinson, London, 1983).

Knower, F.H., Analysis of Some Experimental Variables of Simulated Vocal Expression of the Emotions, *Journal of Social Psychology*, Vol.14 (1941), pp.369–72.

Laing, R.D., Phillipson, H., and Lee, A.R. *Interpersonal Perception* (Tavistock, London, 1966).

Laird, J.D., Self-attribution of Emotion: the Effects of Expressive Behaviour on the Quality of Emotional Experience, *Journal of Personal and Social Psychology*, Vol.29 (1974), pp.475–86.

Larson, Charles V., *Persuasion: Reception & Responsibility* (Wadsworth, San Francisco, 1983).

Lay, C.H., and Burron, B.F., Perception of the Personality of the Hesitant Speaker, *Perceptive and Motor Skills*, Vol.26 (1968), pp.951–6.

Lefebvre, L., Encoding and Decoding of Ingratiation in Modes of Smiling and Gaze, *British Journal of Social and Clinical Psychology*, Vol.14 (1975), pp.33–42.

Lewin, K., Group Discussion and Social Change, in T.M. Newcomb and E.L. Hartley (eds), *Readings in Social Psychology* (Holt, Rinehart and Winston, New York, 1947).

Lewin, Kurt, *Principles of Topological Psychology* (Johnson Reprint Corporation, New York, 1969).

Lewin, Kurt, Edited by Gertrud Weiss Lewin, *Resolving Social Conflicts: Selected Papers on Group Dynamics* (Souvenir Press, London, 1973).

Licklider, J.C.R., and Miller, G.A., The Perception of Speech, in *Handbook of Experimental Psychology*, ed. S.S. Stevens, (Wiley, New York, 1951).

Lindsay, Peter H. and Norman, Donald A., *Human Information Processing: an Introduction to Psychology* (Academic Press, London, 1977).

Lindskold, S., and Tedeschi, J.T., Self-esteem and Sex as Factors Affecting Influencability, *British Journal of Social and Clinical Psychology*, Vol.10 (1971), pp.114–22.

Lorayne, Harry, *The Memory Book* (W.H. Allen, London, 1975).

Lott, R.E., Clark, W., and Altman, I., *A Propositional Inventory of Research on Interpersonal Space* (Naval Research Institute, Washington, 1969).

Luria, A.R., *Working Brain* (Penguin, London, 1976).

McDougall, W., *The Group Mind* (Cambridge University Press, London, 1927).

Maclay, H., and Osgood, C.E., Hesitation Phenomena and Spontaneous English Speech, *Word*, Vol.15 (1959) pp.19–44.

MacManaway, L.A., Using Lecture Scripts, *Universities Quarterly* (June 1968), pp.327–36.

MacManaway, L.A., Teaching Methods in Higher Education – Innovation and Research, *Universities Quarterly*, July 1968, pp.327–36.

Macworth, James, *Vigilance and Habituation* (Penguin, London, 1970).

Mahl, G.F., Disturbances and Silences in the Patient's Speech in Psychotherapy, *Journal of Abnormal and Social Psychology*, Vol.53 (1956), pp.1–15.

Marshall, J.C., The Biology of Communication in Man and Animals, in Lyons, J. (ed.) *New Horizons in Linguistics* (Penguin, London, 1970).

Maslow, Catha, Yoselson, Kathryn, and Harvey London, Persuasiveness of Confidence Expressed via Language and Body Language, *British Journal of Social and Clinical Psychology*, 10 (1971), p.234.

Mastin, V.E., Teacher Education, *Journal of Educational Research*, Vol.56 (1963) pp.385–6.

Mead, Margaret, *Coming of Age in Samoa* (Penguin, London, 1944).

Mehrabian, A., and Diamond, S.G., Seating Arrangement and Conversation, *Sociometry*, Vol.34 (1971) pp.281–9.

Mehrabian, Albert, *Silent Messages* (Wadsworth, San Francisco, 1971).

Mehrabian, Albert, *Non-Verbal Communication* (Aldine Ahterton Inc., Chicago, 1972).

Mehrabian, A., Significance of Posture and Position in the Communication of Attitude and Status Relationships, *Psychology Bulletin*, Vol.1 (1969), pp.359–72.

Mehrabian, A., and Williams, M., Non-verbal Concomitants of Perceived and Intended Persuasiveness, *Journal of Personal and Social Psychology*, Vol.13 (1969), pp.37–58.

Miller, G.A., and Selfridge, J.A., Verbal Context and the Recall of Meaningful Material, *American Journal of Psychology*, Vol.63 (1950), pp.176–85.

Miller, George A., *The Psychology of Communication: Seven Essays* (Basic Books, New York, 1975).

Miller, George A., and Philip Johnson-Laird, *Language and Perception* (Cambridge University Press, Cambridge, 1976).

Miller, George A., and Lenneberg, Elizabeth, (eds), *Psychology and Biology of Language and Thought: Essays in Honor of Eric Lenneberg* (Academic Press, New York, 1978).

Miller, George A., Grusky, Oscar (eds), *The Sociology of Organizations: Basic Studies* (New York Free Press, 1981).

Miller, Gerald R., and Burgoon, Michael, *New Techniques of Persuasion* (Harper and Row, New York, 1973).

Miller, Gerald, R., and Steinberg, Mark, *Between People: A New Analysis of Interpersonal Communication* (Science Research Associates, Chicago, 1975).

Miller, Gerald R., (ed.), *Explorations in Interpersonal Communication* (Sage Publications, Beverley Hills, 1977).

Miller, N.E., and Dollard, J., *Social Learning and Imitation*, (Routledge and

Kegan Paul, London, 1945).

Miller, N., and Campbell, D.T., Recency and Primacy in Persuasion as a Function of the Timing of Speeches and Measurements, *Journal of Abnormal and Social Psychology*, Vol.59 (1959), pp.1–9.

Milroy, Lesley, *Language and Social Networks* (Basil Blackwell, Oxford, 1980).

Mitchell, John, *A Handbook of Technical Communication* (Wadsworth, San Francisco, 1962).

Morris, C.W., *Signs, Language and Behaviour* (Prentice Hall, New York, 1946).

Morris, D., Collett, P., Marsh, P., and O'Shaughnessy, *Gestures, their Origin and Distribution* (Jonathan Cape, London, 1979).

Morris, F.C., *Effective Teaching: A Manual for Engineering Instructors* (McGraw-Hill, Maidenhead, 1950).

Napley, David, *The Technique of Persuasion* (Sweet and Maxwell, London, 1983).

Nuttin, Jozef M., *The Illusion of Attitude Change Towards a Response Contagion Theory of Persuasion* (Academic Press, London, 1975).

O'Donnell, Victoria, and Kable, June, *Persuasion: an Interactive Dependency Approach*, (Penguin, London, 1982).

Packard, Vance, *The Hidden Persuaders* (Pelican, London, 1981).

Paivio, A., Personality and Audience Influence, in B.A. Maher, (ed.), *Progress in Experimental Personality Research* (Academic Press, New York, 1965).

Parker, Paul P., *Develop your Power of Persuasion* (New English Library, 1963).

Patton, Bobby R., and Giffin, Kim, *Interpersonal Communication in Action: Basic Text and Readings* (Harper and Row, London, 1981).

Pearson, Judge and Nelson, Paul B., *Understanding and Sharing* (William Brown, Dubuque, Iowa, 1982).

Pease, Allan, *Body Language: How to Read Others' Thoughts by their Gestures* (Camel Publishing Co., 1984).

Penfield, W., and Perot, P., The Brain's Record of Auditory and Visual Experience: a Final Summary and Discussion, *Brain*, Vol.86, pp.595–702.

Pervin, Lawrence A., *Personality, Theory & Research* (4th edn, Wiley, Chichester, 1984).

Peterson, L.R., Short-term Memory, *Scientific American*, Vol.215, No.1 (July 1966), pp.90–5.

Phares, E. Jerry, *Introduction to Psychology* (Merrill, 1984).

Powell, John, *Why Am I Afraid To Tell You Who I Am? Insights On Self-awareness, Personal Growth and Interpersonal Communication* (Fontana, London, 1975).

Pribram, The Neurophysiology of Remembering, *Scientific American* (January, 1969), pp.73–86.

Quinn, Virginia Nichols, *Applied Psychology* (McGraw Hill, Maidenhead, 1984).

Reardon, Kathleen Kelley, *Persuasion Theory and Context* (Sage, Beverley Hills, 1981).

Reilly, William John, *Opening Closed Minds, And Persuading Others To Act Favourably* (Harper and Row, New York, 1964).

Rejch, Rita Daum, *The Fundamentals of Speech Course: Public Speaking* (Donning Co., Norfolk, Va., 1982).

Reusch, J., and Kees, W., *Non-Verbal Communication* (University of California Press, 1964).

Riesman, D., *The Lonely Crowd* (Yale University Press, New Haven, 1950).

Robinson, Don, and Ray Power, *Spotlight on Communication. A Skills Based Approach* (Pitman, London, 1984).

Robinson, Mike, *Groups* (Wiley, London, 1984).

Robinson, W.P., *Language and Social Behaviour* (Penguin, London, 1974).

Robinson, W.P., (ed.), *Communication in Development* (Academic Press, London, 1981).

Rodman, George R., *Public Speaking: an Introduction to Message Preparation* (Holt, Rinehart and Winston, London, 1981).

Roloff, Michael E. and Miller, Gerald R. (eds), *Persuasion: New Directions in Theory and Research* (Sage Publications, Beverley Hills, 1980).

Roloff, Michael E., *Interpersonal Communication: The Social Exchange Approach* (Sage, Beverley Hills, 1981).

Ross, Raymond S., *Persuasion: Communication and Interpersonal Relations* (Prentice Hall, Englewood Cliffs, 1974).

Ross, Raymond Samuel, *Speech Communication: Fundamentals and Practice* (Prentice-Hall, Englewood Cliffs, 1983).

Ross, Raymond S., *Essentials of Speech Communication* (2nd edn, Prentice Hall, Englewood Cliffs, 1984).

Ross, W.D., (ed.), *The Works of Aristotle*, Vol.11, p.7.

Ruesch, J., and Bateson, G., *Communication, The Social Matrix of Society* (W.W. Norton, New York, 1951).

Ruffner, Michael, *Interpersonal Communication* (Holt, Rinehart and Winston, London, 1981).

Russell, Peter, *The Brain Book* (Routledge and Kegan Paul, London, 1980).

Samovar, Larry A., and Mills, Jack, *Oral Communication: Message and Response* (William Brown, 1967).

Sandell, Rolf, *Linguistic Style and Persuasion* (Academic Press, London, 1977).

Scherer, Klaus R., and Howard Giles, (eds), *Social Markers in Speech* (Cambridge University Press, Cambridge, 1979).

Schlosberg, A., The Description of Facial Expression in Terms of Two Dimensions, *Journal of Experimental Psychology*, Vol.44 (1952), pp.229–37.

Schwerin, Horace S., *Persuasion in Marketing: the Dynamics of Marketing's Great Untapped Resource* (Wiley, Chichester, 1981).

Shannon, Claude and Warren Weaver, *The Mathematical Theory of Communication* (Illinois University Press, 1949), p.34.

Shuter, Robert, *Understanding Misunderstanding: Exploring Interpersonal Communication* (Harper and Row, New York, 1979).

Simons, Herbert W., *Persuasion: Understanding, Practice & Analysis* (Speech Communication Series, Addison Wesley, Reading, MA., 1976).

Skinner, B.F., *Science and Human Behaviour* (Macmillan, London, 1953).

Slater, P.E., Contrasting Correlates of Group Size, *Sociometry*, Vol.21 (1958), pp.129–39.

Slobin, D.I., Miller S.H., and Porter, L.W., Forms of Address and Social Relations in Business Organisation, *Journal of Personal and Social Psychology*, Vol.8 (1968), pp.289–93.

Smart, J.J.C., *Ethics, Persuasion and Truth* (Routledge and Kegan Paul, London, 1984).

Smith, Frederick E., *The Persuaders at Large* (Henry Publications, Brighton, 1976).

Smith, M., Social Situation, Social Behaviour, Social Group, *Psychology Review*, Vol.52 (1945), pp.224–9.

Smithers, A., Some Factors in Lecturing, *Educational Review*, Vol.22 (1970), No.2, pp.141–50.

South, E.B., Some Psychological Aspects of Committee Work, *Journal of Applied Psychology*, Vol.11 (1927), pp.348–68, 437–64.

Sprott, Walter J.H., *Human Groups* (Penguin, London, 1967).

Standing, Lionel, Learning 10,000 Pictures, *Quarterly Journal of Experimental Psychology*, Vol.25, pp.207–22.

Strongman, K.T., and Woosley, J., Stereotyped Reactions to Regional Accents, *British Journal of Social and Clinical Psychology*, Vol.6 (1967), pp.164–67.

Surles, Lynn, *The Art of Persuasive Talking* (McGraw-Hill, New York, 1960).

Tajfel, Henri, and Colin, Fraser (eds), *Introducing Social Psychology*, (Penguin Education, London, 1984).

Thistlethwaite, D.L., and Kamenetzky, J., Attitude Change Through Refutation and Elaboration of Audience Counterarguments, *Journal of Abnormal and Social Psychology*, Vol.51 (1955), pp.3–12.

Thomas, E.J., The Variation of Memory with Time for Information Appearing During a Lecture, *Studies in Adult Education* (April 1972), pp.57–62.

Thompson, Wayne Noel, *The Process of Persuasion Principles and Readings*, (Harper and Row, New York, 1975).

Trenaman, J.M., *Communication and Comprehension* (Longman, London, 1967).

Turner, John, and Giles, Howard (eds), *Intergroup Behaviour* (Blackwell, Oxford, 1981).

University Grants Committee, *Report of the Committee on Audio Visual Aids in Higher Scientific Education* (HMSO, London, 1965).

Vanderlaan, Roger F., *Persuasion* (El Camino, Santa Barbara, CA., 1983).

Verderber, Rudolph F., *Communicate!* (4th edn., Wadsworth, San Francisco, 1984).

Vernon, Gregg, *Human Memory* (Methuen, London, 1975).

Vernon, M.D., *The Psychology of Perception* (Penguin, London, 1982).

Vine, Ian, Judgement of Direction of Gaze, an Interpretation of Discrepant

Results, *British Journal of Social and Clinical Psychology*, Vol.10 (1971), p.320.

Walker, Stephen, *Learning Theory and Behaviour Modification* (Methuen, London, 1984).

Walter, Otis Monroe, *Speaking to Inform and Persuade* (Collier-Macmillan, New York, 1966).

Warr, Peter (ed.), *Psychology at Work* (Penguin, London, 1971).

Warm, Joel, S., (ed.), *Sustained Attention in Human Performance* (Wiley, London, 1984).

Wertheimer, Max, *Productive Thinking* (Tavistock Press, London, 1961).

Whitehouse, Frank, *Systems Documentation Techniques of Persuasion in Large Organizations* (Business Books, London, 1973).

Wicklund, R.A., Objective Self-Awareness, *Advances in Experimental Social Psychology*, Vol.8 (1975), pp.233–75.

Wilkinson, Jill, and Canter, Sandra, *Social Skills Training Manual, Assessment, Programme Design, & Management of Training* (Wiley, London, 1982).

Wilson, John F., and Carroll C. Arnold, *Public Speaking as a Liberal Art* (Allyn and Bacon, Boston, Mass. 5th edn., 1983).

Wilcox, Sidney W., *Technical Communication* (International Text Book Company, Pennsylvania, 1962).

Yates, F.A., *The Art of Memory* (Routledge and Kegan Paul, London, 1966).

Index

OTHER TITLES FROM E & FN SPON

Brain Train
Studying for success
2nd Edition
R. Palmer

Effective Writing
Improving scientific, technical and business communication
2nd Edition
C. Turk and J. Kirkham

Good Style
Writing for science and technology
J. Kirkham

Scientists Must Write
A guide for better writing for scientists, engineers and students
R. Barrass

Study!
A guide to effective study, revision and examination technique
R. Barrass

Studying for Science
A guide to information, communication and study techniques
E.B. White

Write in Style
A guide to good English
R. Palmer

Writing Successfully in Science
M. O'Connor

For more information about these and other titles please contact:
The Promotion Department, E & FN Spon, 2–6 Boundary Row, London, SE1 8HN   Telephone 0171–865–0066.